THE GLP-1 EXIT PLAN

THE GLP-1 EXIT PLAN

Step Forward.
Live Free.
Thrive for Life.

The first step-by-step guide to life after GLP-1s—using personalized nutrition, mindset, and your DNA to stay free for life.

Holli Bradish-Lane

Copyright © 2025 by Holli Bradish-Lane

All rights reserved. No part of this publication may be reproduced, distributed, or transmitted in any form or by any means, including photocopying, recording, or other electronic or mechanical methods, without the prior written permission of the publisher, except in the case of brief quotations embodied in critical reviews and certain other noncommercial uses permitted by copyright law.

For permission requests contact:
Refiner's Forge Publishing
2907 CR 103
Florence, CO 81226
RForgePublishing@gmail.com

Ordering Information:
Quantity sales. Special discounts are available on quantity purchases by corporations, associations, and others. For details, contact the publisher at the address above.

Cover design by Platinumedia
Edited by Susan Keillor

Printed in the United States of America

ISBNs
Hardcover: 979-8-9987212-2-9
Paperback: 979-8-9987212-0-5
eBook: 979-8-9987212-1-2

Disclaimer: This is a work of nonfiction. The information in this book is intended for educational and informational purposes only and is not a substitute for professional medical advice, diagnosis, or treatment. Always seek the advice of your physician or another qualified health provider with any questions you may have regarding a medical condition.

Affiliate Disclosure: This publication may reference affiliate links. I may receive a small commission if you choose to make a purchase through these links, at no additional cost to you. Only products and services aligned with the mission of *The GLP-1 Exit Plan* and supporting optimal health outcomes are recommended.

First edition

From the Physician's Perspective

"This book delivers practical steps to safely taper off GLP-1 medications without sabotaging your progress or your health. It's a must-read for anyone ready to transition with confidence and control."

Pamela Buchanan, MD
TEDx Speaker, Author, and Compassionate Physician
drbstrong.com

As a physician—and someone who has personally prescribed and experienced both the benefits and side effects of GLP-1 medications—I know the question on everyone's mind:

Can I come off GLP-1s and still keep the weight off?

The answer is yes, but not without a plan.

GLP-1 medications have been life-changing for many, but they're not meant for everyone long-term. In my clinical experience, there are several reasons patients may consider tapering off:

- **Persistent side effects** like nausea, fatigue, or GI distress that impact daily life.
- **Achievement of health goals** such as weight loss, improved labs, or remission of prediabetes.
- **Financial limitations or access issues**, especially with ongoing supply challenges.
- And perhaps most importantly, a **desire for metabolic independence**—to maintain success without relying on medication indefinitely.

This book fills a major gap by offering a clear, evidence-informed approach to tapering, while emphasizing lifestyle, support, and long-term sustainability. I'm grateful this resource exists and proud to recommend it.

Advance Praise for *The GLP-1 Exit Plan*

"The ultimate toolkit for breaking free from GLP-1s and stepping into a future of empowered, DNA-based well-being."

R. Lyman Lane
Facilities and Operations Leader

***

"A powerful guide to lasting success, resilience, and true health."

Dale Amory
Owner, Imagine Yoga and Sound Studio
imagineyogastkitts.com

***

"An amazing whole-person approach to health. Holli Bradish-Lane moves beyond the scale, offering a truly health-minded strategy that will be an invaluable tool for so many."

Jennille Spellman
Holistic Health Advocate

***

"Very inspiring and packed with foundational wisdom for living a healthier, stronger life. As someone who personally takes these medications — and knowing several friends who do as well — I can say the information shared here goes far beyond anything we've ever received from our doctors. This is an amazing resource and one that fills a crucial gap in care."

The book is incredibly well-organized, with clear transitions between topics that build naturally on each other. It makes complex information easy to follow and apply.

I especially loved the discussion questions and exercises at the end of each chapter. They are thought-provoking and will ensure readers walk away not just understanding the material-- but knowing exactly how to put it into practice in their own lives.

I'm so impressed with the obvious care and hard work behind this project. Every part of it was a joy to read, and I know it will give hope to so many — whether they are worried about transitioning off medications or looking for empowering alternatives. An outstanding and much-needed guide!"

Susan Keillor
Author, Certified Life Coach, and Editorial Consultant

***

"Holli Bradish-Lane offers a vast array of proven weight loss tools and strategies in The GLP-1 Exit Plan. As a celiac, I have had a complicated relationship with carbohydrates for over 30 years. Traditional weight loss plans complicated the situation, with reliance on low-carb gluten-containing options that were not workable for me. With revelations from my DNA testing, I was able to incorporate carbohydrates in my diet in a systematic way that did not leave me craving basics.

The GLP-1 Exit Plan is a mix of academic study and personal observations, part literature review, introspection, and a pinch of self-deprecation. Weight loss is extremely context-specific, including the fact that our journeys may have been historically unpredictable and sometimes chaotic. With so many weight loss options out there, The GLP-1 Exit Plan provides science-based guidance to begin, implement, and maintain a strategy that includes overcoming hurdles."

Janice Lowstuter
Regulatory Nurse Executive

***

"This isn't just another diet book—it's a game-changer. Holli combines cutting-edge science, mindset shifts, and actionable steps to create a revolutionary approach to health. As a leading expert in DNA-based health coaching and a masterful storyteller, she presents complex concepts in a way that's down-to-earth, engaging, and easy to understand.

If you've ever wondered how your genetics impact weight loss and overall well-being, this book is your answer. Holli seamlessly blends personal experiences with professional expertise, offering insights that are both relatable and deeply informative.

More than just a guide—it's a purpose-driven roadmap to a healthier life. Ready to take control of your health? Start here."

Alan D. King
Retired Public Servant

***

"A smart guide to real, lasting health."

Joyce Coffey
Pharmacy Leader

For those who chose to become.

You let go. You stepped forward. You rose.

This is for you.

Contents

Preface xvii
A Note Before You Begin xix
Important Considerations for Your Health
How to Use This Book for Maximum Results

Honoring Privacy, Sharing Truth xx

Part 1: Understanding the Journey 1

Introduction: Your Journey Starts Here 3
From GLP-1s to Lifelong Wellness
Who This Book Is For
You're Not Broken — You're Just Getting Started

Chapter 1: Understanding GLP-1 Medications and Their Role in Weight Loss 9
How They Work, Why They Help, and When to Transition

Chapter 2: Understanding GLP-1s— And Why You Don't Need Them Forever 25
What They Don't Do — and What Your Body Can
The Power of Personalization and the DNA Factor

Part 2: Reclaiming Your Mindset and Empowerment 37

Chapter 3: The Shift That Changes Everything— Break Free from the Cycle of Self-Doubt 39
Letting Go of Limiting Beliefs

Chapter 4: Shifting Your Mindset for Success— Rewiring Your Thinking to Thrive Beyond GLP-1s 49
Emotional Eating, Self-Sabotage, and 1% Better

Chapter 5: The 7 Steps to Lasting Transformation with Neuro-Associative Conditioning (NAC) (Rewiring Your Brain for Sustainable Change) 77
Rewiring Habits with Neuro-Associative Conditioning

Part 3: Personalizing Your Plan for Lasting Results 93

Chapter 6: The DNA Connection: Unlocking Your Body's Blueprint for Sustainable Weight Loss 95
How Genetics Shape Weight, Nutrition, and Exercise

Chapter 7: Rewiring Your Neuro-Associations for Freedom 117
The Tomorrow Technique & Managing Cravings

Chapter 8: The Power of Belief and Certainty— Trusting Yourself to Achieve What You Truly Want 133
How Mindset Drives Biology and Behavior

Part 4: Sustainable Habits for Whole-BEING Wellness 149

Chapter 9: Building Your Nutrition Foundation— The Cornerstone of Lasting Health and Vitality 151
Eating for Your Genetics & Balancing Hunger

Chapter 10: Fitness for Your DNA and Lifestyle 173
Personalized Movement for Energy and Longevity

Chapter 11: Stress, Sleep, and Whole-BEING Wellness 191
Practical Tools for Resilience and Metabolic Health

Part 5: Thriving for Life 213

Chapter 12: Breaking Free—Strategies for Tapering Off GLP-1 Medications 217
How to Know When You're Ready and What to Expect

Chapter 13: Your Long-Term Plan— Thriving Without GLP-1 Medications 245
Handling Setbacks and Staying Empowered

Conclusion: Empowered for Life 261
Sustainable Habits, Lifelong Wellness, and What's Next

Resources 265
Workbook, Coaching, Tools & Community

About the Author 273
Final Thoughts, Stay Connected, and Keep Thriving 275

Preface

This book wasn't born overnight. It came from lived experience, tough questions, and the quiet, often invisible work of rebuilding from the inside out. And if you're here, holding these pages, chances are you've carried a lot — for a long time.

You may have used medication to support your journey.

You may have done everything "right" — tracked, fasted, restricted, stuck to the plan — chasing the promise of change that always seemed just out of reach.

You might be tired of starting over, wondering if change that lasts is possible.

It is.

Maybe somewhere along the way, you started wondering if the way you've been doing strength is sustainable. That wondering is sacred. It's not weakness. It's wisdom.

This book isn't about pushing harder. It's about rebuilding differently. It's about recognizing the refining process you're already in — and realizing you're not alone in it.

I wrote this for the quiet rebuilders.

The ones who are choosing to become.

If you're ready to approach your health differently, you're not alone — and you're exactly where you need to be.

A Note Before You Begin

Before we begin, it is important to recognize that this book is not a substitute for professional medical advice, diagnosis, or treatment. The insights, strategies, and recommendations presented here are designed to empower you with tools and knowledge for your health journey, but they are not a replacement for the guidance of your healthcare provider.

If you are currently taking GLP-1 medications (such as Wegovy, Ozempic, Mounjaro) and are considering discontinuing them, please consult your prescribing physician before making any changes to your treatment plan. GLP-1 medications are powerful tools for weight loss, and tapering off of them requires a thoughtful, personalized strategy. **This book serves as a resource to help you better understand the process, identify actionable steps, and add meaningful context to the plan you create in collaboration with your healthcare provider.**

The process of tapering off of GLP-1 medications is somewhat uncharted territory. While these medications are effective for weight loss, they are not a complete solution for long-term weight management. This book aims to fill the gap by offering practical strategies rooted in science, my personal experience, and the successes of the DNAslim: Whole-BEING program. It will empower you with tools to build sustainable habits, address hunger and cravings, and optimize your health beyond these medications.

As a licensed healthcare clinician, I am deeply committed to helping individuals achieve lasting results. However, every person's journey is unique, and the results described in this book will vary based on individual circumstances, genetics, and adherence to the strategies outlined here. While this book is informed by my professional expertise and research, it is not a replacement for personalized medical care.

The tools in this book are designed to complement your existing medical care and support your long-term health goals. You deserve an approach that works for you, empowers you, and helps you navigate this exciting next chapter with clarity and confidence.

Honoring Privacy, Sharing Truth

Throughout this book, I share real stories of transformation—people who have reclaimed their health, confidence, and freedom. To respect their privacy, names and identifying details have been changed, but their experiences remain true. In full transparency, my personal stories are shared as they truly happened.

Now, let's get started on your journey toward lifelong health, wellness, and whole-BEING empowerment!

Part 1

Understanding the Journey

Breaking Free from Quick Fixes and Building a Strong Foundation

"The beginning is always today."

— Mary Shelley

You're here because you want more than a temporary solution. You want lasting change. Part One is all about understanding how your body, mindset, and metabolism work—so you can finally break free from the cycle of frustration.

We'll start by exploring GLP-1 medications—how they work, why they help, and why they were never meant to be a lifelong crutch. More importantly, you will see why *you* were always the one in control—not the medication.

You'll also begin shifting your mindset from relying on external solutions to building internal strength. This is the first step toward lasting transformation—understanding that success isn't about *what* you take, but *who* you become.

By the end of this section, you will have a clear roadmap for your journey and the knowledge to move forward with confidence. Let's get started.

Introduction

Your Journey Starts Here

From GLP-1s to Lifelong Wellness: Unlock Your Body's True Potential

I've been where you are.

Not just physically—but emotionally, mentally, and spiritually, too.

For years, I carried more than just extra weight—I carried the weight of exhaustion, shame, and silent battles that no one could see. My life, on the surface, might have looked successful. I held high-stress leadership positions that demanded my full attention, where I was praised for my ability to juggle responsibilities and "get things done." But beneath the polished exterior, I was crumbling. I was overworked, overwhelmed, and completely out of touch with myself.

I remember the way my days would blur together: the endless meetings, the relentless deadlines, and the constant feeling that I was chasing a finish line that kept moving further out of reach. My evenings were no better—I'd collapse on the couch, too drained to cook, let alone exercise. My body felt heavy, my energy was nonexistent, and my mind was clouded with doubt.

At home, things weren't much easier. I found myself stuck in an unhealthy relationship, one that quietly chipped away at my sense of self-worth. I didn't feel supported, understood, or truly seen. And in that loneliness, I turned to food for comfort—a bowl of ice cream here, a bag of chips there. For a fleeting moment, it would feel like relief. But afterward, the guilt would hit, piling onto the mountain of shame I was already carrying.

The physical toll was inevitable.

First, there was the inflammation. I felt it in my joints, a constant ache that made even simple tasks—like climbing the stairs—feel monumental. Then came the weight gain, creeping up slowly at first, until it became impossible to ignore. The person I saw in the mirror didn't feel like me anymore. I didn't recognize her. And the more I tried to "fix" it—jumping into the latest diet trends, logging every calorie, pushing through painful workouts—the more my body seemed to resist.

As if that wasn't enough, life kept throwing curveballs.

I went through seven knee surgeries, each one leaving me more frustrated and less mobile than before. I was faced with infertility; a deeply personal battle that made me feel like my body was betraying me in every way. And then, the ultimate scare: a lump in my breast that sent my world spinning. I waited for the test results, grappling with the possibility of breast cancer and all the fears that came with it.

It was too much.

I felt like my body had become my enemy. It was as though no matter what I did—no matter how hard I fought—it just kept breaking down, demanding more than I had to give. I was stuck in a spiral of exhaustion, self-doubt, and hopelessness. And worst of all, I felt like I was failing—not just myself, but everyone who relied on me.

Through it all, I prayed. I held onto my faith, though at times it felt fragile. I had always believed that God had a plan for my life, that He worked all things for good, that I wouldn't be given more than I could handle—but this? This felt insurmountable. In those darkest moments, I wondered if He had forgotten me.

But here's the thing about hitting rock bottom: It gives you nowhere to go but up.

The Turning Point: My Personal Journey

I will never forget the moment when everything started to shift. I had spent years struggling with my weight and inflammation, trying numerous conventional approaches without lasting success. Frustrated and searching for real answers, I came across something that changed everything—a program that used genetics to personalize health plans. At first, I dismissed it. It sounded too good to be true—like another gimmick or quick fix. But later that night, as I lay awake staring at the ceiling, I found myself wondering: *What if?*

I started researching the science behind genetics and weight loss, and what I discovered blew me away. I learned that our DNA plays a significant role in how our bodies process food, store fat, and respond to exercise. For example, certain genetic markers—like the FTO gene—can make us more prone to hunger, while others influence how well we metabolize carbs or fats. Suddenly, all those years of frustration started to make sense.

I realized that it wasn't a lack of willpower holding me back. It wasn't that I was lazy or broken. The problem was that I'd been trying to follow plans that weren't designed for my body in the first place.

So, I decided to take the leap.

I ordered a DNA test, sent in my sample, and waited. When my results came back, it was like a lightbulb went off. I finally had answers: why certain diets had failed me, why some exercises left me feeling depleted instead of energized, and why my cravings seemed impossible to control. Armed with this knowledge, I began creating a plan tailored to my body's unique blueprint.

But I also knew that genetics were just one piece of the puzzle. To truly transform, I needed to address the bigger picture: my mindset, my habits, and my relationship with myself. It wasn't just about weight loss anymore—it was about becoming whole again. It was about building a life that felt sustainable, joyful, and empowering.

I started to understand that my struggles had not been punishments, but preparation. That the roadblocks I faced weren't meant to break me, but to redirect me toward something greater. And through it all, I saw the hand of God. I wasn't just learning how to heal my body—I was learning how to trust the process, to have faith in the journey, and to believe that even in the hardest moments, I was being led somewhere purposeful.

That journey is what inspired me to create DNAslim—a program that combines the power of genetic insights with proven strategies for long-term health and wellness. It's not just a weight-loss program; it's a complete whole-BEING transformation, designed to help you unlock your full potential and thrive.

And if you're here, reading this, I believe you are meant to step into that transformation, too.

No matter how many times you've struggled, no matter how hopeless it has felt, your story isn't over. There is a path forward—one that honors your body, your mind, and your spirit.

It's time to step into the healthiest, strongest, most empowered version of you.

Why This Book? Beyond Quick Fixes

Over the years, I've worked with many people who felt like I did: stuck, frustrated, and hopeless. Many of them turned to GLP-1 medications—like Wegovy, Ozempic, or Mounjaro—as a lifeline. And for good reason: these medications are game changers for weight loss. They can reduce hunger, improve blood sugar control, and help you lose significant amounts of weight.

But what happens when the prescription runs out?

The truth is, GLP-1 medications are a tool—not a solution. Without a plan in place, people find themselves right back where they started. Studies show that individuals coming off these medications regain two-thirds of the weight they lost within six months. And it's not because they failed. It's because medications can't replace the habits, mindset, and lifestyle changes needed for long-term success.

That's where this book comes in.

Whether you're still on your GLP-1 medication journey, working toward your ideal weight, or you've reached your goal and are transitioning into maintenance, this book is your bridge—a step-by-step guide to building a foundation for lifelong health. It's about equipping you with the tools, knowledge, and confidence to take control of your health without feeling dependent on medication.

And if you're not currently taking GLP-1s but are considering them, this book is for you, too. Before starting, why not try a strategy designed to work with your body's natural biology? You may find that with the right approach, you can achieve lasting results without ever needing the medication.

Who This Book Is For

This book is for you if:

- You're currently using GLP-1 medications and want to taper off of them successfully—whether you're still on your weight-loss journey or in maintenance mode.
- You're tired of diets that feel restrictive or unsustainable.
- You want to understand how your body works and how to create a personalized plan that aligns with your unique needs.

- You're ready to stop feeling stuck and start building a life of energy, confidence, and empowerment.
- You've experienced side effects or challenges with GLP-1 medications and want an alternative path that supports your body and long-term health.

This isn't just another diet book. It's a revolutionary approach that combines science, mindset, and action. Inside, you'll learn:

- How to taper off of GLP-1 medications without regaining weight.
- How to use **your DNA Blueprint** (or self-discovery tools provided in the book) to personalize your health journey.
- How to manage hunger, reduce cravings, and fuel your metabolism with the right foods.
- How to build habits that last a lifetime, without feeling restricted or overwhelmed.
- How to address the emotional and psychological barriers that hold you back, so you can feel empowered in every aspect of your life.

You're Not Broken. You're Just Getting Started

I know this journey isn't easy. But I also know that it's possible—because I've lived it. I've seen the power of transformation, not just in myself, but in the people I've had the privilege to coach.

You're not alone. And you're not broken. You're about to unlock the tools to create the life you deserve—one that's full of energy, confidence, and resilience.

This isn't just about weight loss. It's about becoming the strongest, healthiest, and most empowered version of yourself.

Let's begin.

Chapter 1

Understanding GLP-1 Medications and Their Role in Weight Loss

What Are GLP-1 Medications?

For many people, GLP-1 medications like **Wegovy**, **Ozempic**, and **Mounjaro** feel like the answer they've been searching for. After years of battling hunger, cravings, and the emotional toll of weight gain, these medications can seem like a breakthrough. For the first time, food doesn't feel like it has control over you—you feel in control.

But what exactly are GLP-1 medications, and why do they work so well?

Let's start with the science. GLP-1 stands for **glucagon-like peptide-1**, a hormone your body naturally produces in your gut. It plays a vital role in regulating appetite, blood sugar, and digestion. When you eat, GLP-1 is released to signal fullness, slow down how quickly your stomach empties, and help your body use insulin more efficiently.

GLP-1 medications—technically known as **GLP-1 receptor agonists**—mimic and enhance the effects of this natural hormone. For simplicity, we'll refer to them as GLP-1 medications throughout this book. Here's how they work:

- **Suppressing Hunger**: GLP-1s reduce the hormones that trigger hunger and amplify the ones that signal fullness.
- **Controlling Cravings**: By stabilizing blood sugar, they reduce the highs and crashes that often lead to intense cravings.

- **Slowing Digestion**: By delaying gastric emptying, they help you feel fuller for longer after meals.

Originally developed to treat **type 2 diabetes**, GLP-1s were later found to produce significant weight loss. Clinical trials showed participants lost an average of **15–20% of their body weight**—far more than what most people achieve through diet and exercise alone.

For those struggling with obesity or weight-related health conditions, this was a game changer. These medications don't just suppress appetite—they directly address the **biological drivers of weight gain**.

But as powerful as they are, GLP-1 medications are **tools**, not cures. And the long-term results depend on how—and whether—you build the foundation to thrive beyond them.

Why GLP-1s Are So Effective for Weight Loss

If you've ever tried dieting, you know this: **hunger is relentless**.

You eat less. You cut calories. You try your best to follow the plan. But eventually, that gnawing, primal hunger shows up—and it's stronger than motivation.

This is where GLP-1 medications truly shift the game.

They help **quiet hunger signals**, reduce cravings, and balance blood sugar in a way that feels like your body is finally cooperating. You feel full on less food. You have fewer cravings. Your energy stays more stable. Most importantly—you're not in a constant battle with your biology.

GLP-1s also improve **insulin sensitivity**, which can be a game-changer for individuals with prediabetes, PCOS, or type 2 diabetes. Improved insulin sensitivity means your body is better at managing glucose, storing less fat, and experiencing fewer crashes that trigger binge cycles.

And while weight loss is a major benefit, many users also report an emotional shift:

- **Less mental chatter about food**
- **Less shame around eating**
- **More freedom and peace in their relationship with their body**

These medications don't just help you lose weight. They help you regain trust in yourself.

Who Can Benefit from GLP-1s?

GLP-1 medications aren't for everyone—but for the right person, they can be transformative.

They're typically prescribed for:

1. **Obesity:** BMI of 30 or higher
2. **Overweight with Comorbidities:** BMI of 27 or higher and health conditions like high blood pressure, high cholesterol, or type 2 diabetes
3. **Type 2 Diabetes:** GLP-1s remain a first-line treatment for blood sugar control

But this isn't just about numbers. Many people turn to GLP-1s after years of yo-yo dieting, metabolic burnout, or hormone-related weight gain. For them, these medications feel like a way back to themselves.

Why People Turn to GLP-1 Medications

For many, GLP-1 medications represent the first time their **body feels like an ally**, not an obstacle. Hunger fades. Cravings lessen. Motivation returns. There's space to think clearly, eat mindfully, and feel hope again.

Women make up a significant percentage of GLP-1 users. Societal pressure, hormonal shifts like menopause or postpartum changes, and decades of toxic diet culture have pushed many women to seek lasting solutions. For them, GLP-1s are more than medication—they're a lifeline.

These medications offer something deeply valuable: **relief**. Relief from the mental and emotional fatigue of fighting food, hunger, and shame.

The Risks and Unknowns of GLP-1 Medications

While GLP-1 medications are effective, they're not without risks—and some of their long-term effects are still being studied.

Vision Issues

Some users report changes in vision, including blurred vision or worsening of diabetic retinopathy. A study published in *JAMA Ophthalmology* linked GLP-1 medications—particularly semaglutide—to an increased risk of a rare but serious condition called **non-arteritic anterior ischemic optic neuropathy (NAION)**, which can lead to sudden vision loss (1).

- People with diabetes were over **4x more likely** to develop NAION
- People with obesity had over **7x the risk** compared to non-users

While rare, it's a reminder to monitor eye health closely, especially for those with a history of vision conditions.

Gastrointestinal Side Effects

Nausea, vomiting, diarrhea, and constipation are the most common side effects. For most people, these symptoms improve over time—but for others, they can be persistent and disruptive (2).

A more severe condition, **gastroparesis** (delayed gastric emptying), occurs when food stays in the stomach longer than normal. This can lead to:

- Bloating and early satiety
- Acid reflux and indigestion
- Vomiting undigested food hours after meals

While slowed digestion is part of how GLP-1s work, in some individuals, this effect becomes extreme and doesn't resolve over time.

GLP-1s have also been linked to **gallstones** and **gallbladder inflammation**, particularly during periods of rapid weight loss. Some people have required gallbladder removal after prolonged use.

Thyroid Concerns and the Black Box Warning

GLP-1 receptor agonists carry a **Black Box Warning**—the most serious safety alert issued by the U.S. Food and Drug Administration (FDA).

What is a Black Box Warning?

It highlights **severe or life-threatening risks** associated with a medication and appears prominently on the label to alert healthcare professionals and patients.

In the case of GLP-1s, animal studies showed an increased risk of **thyroid C-cell tumors**, including **medullary thyroid carcinoma (MTC)**. While human studies have not confirmed the same risk, people with a personal or family history of MTC or **multiple endocrine neoplasia syndrome type 2 (MEN 2)** are advised to avoid these medications (3).

Other Emerging Concerns

1. Muscle Loss and Nutrient Deficiencies

GLP-1s can promote significant weight loss—but not all of it comes from fat. Without adequate protein and resistance training, **lean muscle mass** can be lost. This can impact metabolism, energy, and long-term physical resilience.

Also, due to reduced appetite and slowed digestion, some users may develop **nutrient deficiencies**, especially in:

- Protein
- Iron
- B-vitamins
- Fat-soluble vitamins (A, D, E, and K)

2. Mood Changes and Emotional Blunting

Some individuals report increased anxiety, depression, or a sense of emotional flatness. This may be related to changes in **dopamine signaling**, which influences mood, motivation, and pleasure. For some, even a lack of interest in food spills over into a general disconnection from joy.

3. Increased Heart Rate (Tachycardia)

Mild but persistent increases in **resting heart rate** have been observed in some users. While the long-term impact is still unclear, those with pre-existing cardiovascular concerns should monitor heart rate regularly.

4. Pancreatitis

Rare but serious, pancreatitis can occur during GLP-1 use. Symptoms include **persistent abdominal pain**, nausea, and vomiting. Seek immediate medical attention if these arise.

5. Hair Loss (Telogen Effluvium)

While not caused directly by GLP-1s, some users report hair shedding, often **2–3 months into use**. This may be linked to the metabolic stress associated with rapid weight loss, though some individuals have experienced hair loss even in the absence of weight loss. Ensuring proper **protein, iron, and micronutrient intake** may help mitigate this effect.

Long-Term Effects

Because GLP-1 medications are relatively new in the weight-loss space, their long-term safety is still under investigation. While short-term clinical trials have demonstrated efficacy, concerns remain around:

- **Pancreatic health** (due to rare reports of pancreatitis and changes in pancreatic enzymes)
- **Gallbladder function** (including gallstones and inflammation)
- **Bone metabolism** (some studies suggest a possible link between rapid weight loss and reduced bone density)

That doesn't mean these medications are unsafe. But it *does* mean that continued monitoring and **informed decision-making** are key.

The goal is not to scare you—it's to **equip you**. You deserve the full picture so you can make empowered choices for your long-term health.

GLP-1s and the Unspoken Side Effects: Intimacy, Desire, and Connection

Most conversations about GLP-1s focus on weight loss, blood sugar, or digestive side effects. But what about the **parts of life that don't show up on lab work?** What about intimacy, connection, and pleasure?

Some users report shifts in **libido**, changes in **sexual desire**, or challenges with physical connection. While the research is still emerging, early evidence suggests GLP-1 medications may influence **dopamine signaling**—the neurotransmitter tied to motivation and pleasure. For some, this may lead to **emotional dullness** or a drop in sexual interest. For others, improved energy and body image may **enhance** confidence and intimacy.

There are also contradictory findings related to **erectile function**. Some men experience improvements due to better circulation and metabolic health. Others report temporary challenges—possibly due to hormonal shifts or decreased dopamine sensitivity.

As with all things in health, **your experience is personal and valid.** These changes may not show up in the clinical literature yet—but if you're noticing a shift in desire or connection, you're not imagining it. And you're not alone.

The solution? Awareness. Open dialogue. And a holistic plan that supports not just weight loss—but **whole-BEING wellness.**

That includes:

- Optimizing nutrition and micronutrient intake
- Supporting hormone balance naturally
- Stress reduction and sleep optimization
- Open communication with your partner and provider

Your vitality, intimacy, and connection matter. This journey isn't just about pounds lost—it's about feeling fully alive, **inside and out**.

Why Do People Want to Taper Off GLP-1s?

As effective as GLP-1s can be, many people reach a point where they start asking: *What's next?*

Here are the most common reasons people consider tapering off:

1. Cost

Without insurance, GLP-1 medications can cost between **$1,000 and $1,500 per month**. Even with coverage, co-pays and deductibles can add up quickly. For many, the financial burden becomes unsustainable long term.

2. Side Effects

While many users tolerate the medication well, others experience **persistent GI issues, fatigue, dizziness, or gallbladder complications**. These side effects—especially when they impact daily quality of life—can lead people to reconsider.

3. Accessibility and Shortages

As demand increases, many have experienced **supply chain issues** or pharmacy delays. Others have been forced to switch brands or reduce dosages due to back-orders—adding to stress and uncertainty.

4. Desire for Long-Term Freedom

Many see GLP-1s as a **kick-start**, not a lifelong plan. There's a growing desire to take ownership of health **without medication dependency**, to know that their body can maintain success through *strength, strategy, and science*—not just prescriptions.

5. Fear of Dependency

This is one of the most emotionally charged concerns:

- *Will I regain the weight if I stop?*
- *Will the hunger and cravings come back stronger than before?*
- *Will all my progress disappear?*

Studies show that people can regain **up to two-thirds** of the weight lost within six to 12 months of stopping GLP-1s—especially if no long-term strategy is in place. But this isn't a reflection of failure. It's biology.

When you remove a medication that regulates appetite and insulin, your body naturally tries to **return to its previous set point**. But here's the good news:

With the right plan—one that honors your biology, supports your metabolism, and rewires your mindset—you can maintain your progress.

And that's exactly what this book is here to help you do.

The Role of GLP-1s in the Bigger Picture: Tools, Not the Only Solution

GLP-1 medications have opened doors for millions. They've helped people regain confidence, reduce disease risk, and reconnect with their bodies. That matters.

But it's also important to understand the **context**.

GLP-1s are part of a multi-billion-dollar pharmaceutical industry. These medications are marketed as long-term solutions—sometimes without equal emphasis on the role of **habits, mindset, or personalized nutrition**.

They're powerful tools. But they don't address the **root causes** of weight challenges:

- Emotional eating
- Blood sugar instability
- Inflammation
- Poor sleep
- Stress
- Nutrient imbalances
- Lack of metabolic muscle

They also don't consider your **unique DNA**—the genetic blueprint that influences how your body responds to food, exercise, hunger, and even motivation.

That's why GLP-1s, while transformative, are not the *only* solution—and certainly not the *final* one.

The true solution is **you**, equipped with personalized knowledge, strategic support, and tools designed for your biology—not someone else's.

Why It's Important to Decide What's Right for You

This journey is personal.

For some, staying on GLP-1s long-term may be the best choice—especially for those managing type 2 diabetes or other metabolic disorders. These medications can be part of a safe, effective maintenance plan when monitored correctly.

For others, the idea of staying on medication forever feels limiting—physically, emotionally, or financially.

And for many, the desire to taper off comes from a deeper place:

A longing to **trust their body again.** To know they can eat, move, and live with **freedom, clarity, and strength**—without needing a prescription to feel "in control."

There is no one-size-fits-all answer. That's why this book exists. To help you decide what's right for *you*, based on:

- Your health history
- Your values
- Your goals
- Your **DNA Blueprint**
- And the life you want to live

Your Opportunity for Freedom

The word "freedom" means different things to different people.

For some, it's **freedom from cravings.**

For others, it's **freedom from medication costs** or the fear of relapse.

And for many, it's **the freedom to finally trust themselves.**

This book is your roadmap to that freedom.

Yes, GLP-1s may have helped you get started. But they're not the destination. The destination is a lifestyle built on strength, clarity, strategy, and self-leadership. It's built on the tools you'll develop in the pages ahead—nutrition that matches your DNA, habits that rewire your brain, and movement that honors your body.

You've already proven you're capable of change. Now it's time to create **lasting transformation.**

Looking Ahead: Beyond the Medication—Building Lasting Strength and Freedom

Tapering off GLP-1 medications can feel like unsteady ground. Hunger may return. Cravings may reappear. Doubt might creep in.

But you are **not starting over.**

You are stepping into a new chapter—one where your tools go deeper, your strategies are more personal, and your success is driven by **empowerment, not medication.**

The path ahead includes:

- Stabilizing hunger naturally
- Balancing your blood sugar with food timing and DNA-informed nutrition
- Rebuilding muscle to fuel metabolism
- Reclaiming mindset and belief
- Developing **neuro-associative patterns** that end self-sabotage
- And learning how to live in a body you respect—without relying on a weekly injection

The Opportunity

Here's the truth:

GLP-1 medications are powerful tools. But **you are more powerful.**

This book will guide you step-by-step through your transition: from medication dependency to metabolic freedom. You'll create a lifestyle grounded in **your DNA**, your vision, and your ability to adapt for life.

You are not broken. You're just getting started.

Up Next

In the next chapter, we'll explore the **mindset shifts** that create real transformation—whether you're still using GLP-1s or stepping into life beyond them.

Your body is ready.

Your mind is next.

Let's go.

References

1. Early worsening of diabetic retinopathy in people treated with semaglutide: Bain SC, et al. *Diabetes Obes Metab.* 2019;21(5):920-926. DOI:10.1111/dom.1362
2. Novo Nordisk. (2023). *Potential vision effects in GLP-1 users.*
3. Gastrointestinal adverse effects of GLP-1 receptor agonists: Abd El Aziz M, et al. *Therap Adv Gastroenterol.* 2020;13:1756284820929308. DOI:10.1177/1756284820929308
4. Journal of Endocrinology and Metabolism. (2022). *Long-term safety of GLP-1 receptor agonists: A review.*
5. Thyroid C-cell tumors and GLP-1 receptor agonists: Marso SP, et al. *N Engl J Med.* 2016;375:311-322. DOI:10.1056/NEJMoa1603827
6. GLP-1 receptor agonists and pancreatitis/gallbladder disease risk: Monami M, et al. *Diabetes Res Clin Pract.* 2017;128:24-34. DOI:10.1016/j.diabres.2017.04.013

Putting Chapter 1 into Action

Your Turn: Integrate & Empower

Understanding GLP-1 medications is just the first step—lasting success comes from building habits that support your body, whether or not medication is part of your journey. These quick exercises will help you reflect on your progress, strengthen your awareness around hunger and cravings, and set a solid foundation for the road ahead.

Take a few moments to complete these mini exercises now, and remember: small, intentional steps lead to lasting transformation. You may want to use a journal for written reflection.

- Want to go deeper? A link to the full Whole-BEING Empowerment Workbook is available in the Resources section of this book. There, you'll find expanded exercises, action steps, and reflection prompts to personalize your journey.

Mini Exercise: Understanding Your Health Journey and Hunger Awareness

1. Your Health Journey Reflection

Take a moment to reflect on where you've been and where you want to go.

- What has your weight loss journey looked like so far?
- What has worked well for you, and what hasn't?
- What does success look like for you beyond just the scale?

📌 *Action Step:* Write a few sentences about what you want to achieve—not just in terms of weight, but in how you want to **feel** in your body and daily life.

2. Hunger & Cravings Awareness

- When do you feel true physical hunger (stomach growling, low energy) vs. when do you experience cravings (boredom, stress, emotional triggers)?
- What are your most common craving triggers?

📌 *Action Step:* At your next meal, **slow down** and check in with yourself halfway through. Are you still physically hungry, or are you eating out of habit?

3. Set Your Baseline Habits

To build a strong foundation, start with two simple habits that support metabolism, energy, and overall well-being:

1. **Move Daily:** Walk for **15 minutes** each day as a time to reflect and reset.
2. **Hydrate Well:** Aim for **1 oz of water per kg of body weight** (*Body weight in lbs. ÷ 2.2 = kg*).

📌 *Action Step:* Calculate your daily water goal and track your intake for the next few days.

Chapter 2

Understanding GLP-1s— And Why You Don't Need Them Forever

*Why Understanding the Science
Puts You in Control of Your Health Journey*

Beyond the Lifeline: What Happens Next

Picture this: You've been using GLP-1 medications for months now. You've seen changes in your body that have felt out of reach for years. The hunger has quieted. The cravings no longer feel like a tidal wave. And for the first time, you've started to believe: *Maybe I can actually do this.*

But now, you're starting to wonder: *What happens next?*

Whether you've just started your GLP-1 medication journey, or you've already seen success, it's normal to ask questions. Can I maintain this progress without the medication? Will my hunger come roaring back? Will I regain the weight?

You're not alone in these concerns, and they're valid. Studies have shown that many people who stop GLP-1 medications regain a significant amount of the weight they lost. That might sound scary, but here's the truth: It's not because you've failed. It's because GLP-1s are only one part of the puzzle.

This chapter will help you understand exactly what GLP-1 medications are doing in your body and, more importantly, why your long-term success doesn't depend on staying on them forever. By understanding the science, you'll empower yourself to take control and build sustainable habits that allow you to taper with confidence.

Recap: How GLP-1s Work

In the last chapter, we explored how GLP-1 medications help regulate appetite, cravings, digestion, and insulin sensitivity. Here's a quick review of their key effects:

1. **Reducing Hunger** – GLP-1s signal your brain's appetite center, helping you feel full faster and stay satisfied longer.
2. **Controlling Cravings** – By stabilizing blood sugar, GLP-1s reduce the intense cravings for sugar and processed foods that can derail progress.
3. **Slowing Digestion** – These medications delay how quickly food leaves your stomach, naturally decreasing the urge to overeat.
4. **Improving Insulin Sensitivity** – For those with insulin resistance, GLP-1s help the body use insulin more effectively, making weight loss more sustainable.

Understanding how GLP-1s work is empowering because it allows you to prepare for what happens next. These medications aren't magic—they simply help regulate systems that may have been out of balance. But here's the good news: Your body is adaptable. With the right strategies, you can train your metabolism, appetite signals, and habits to work in your favor— without relying on medication forever.

That's what this book is here to help you do. In the next sections, we'll go deeper into how you can maintain these benefits naturally, support your body's own GLP-1 production, and build the kind of lifestyle that sustains your results for life.

What GLP-1 Medications Don't Do—And Why That Matters

As powerful as GLP-1 medications are, they have their limits. They can quiet hunger, reduce cravings, and help regulate blood sugar—but they don't address the **root causes** of weight struggles.

They don't:

- Build healthy habits that last a lifetime.
- Teach you how to respond to triggers or emotional eating.
- Help you rewire your relationship with food.
- Address the psychological and behavioral factors that lead to weight gain.

GLP-1 medications are tools, not solutions. They can help you take the first steps, but they can't walk the entire journey for you.

And here's the thing: If you only rely on the medication without building a strong foundation, transitioning off GLP-1s can feel like losing your safety net. Hunger may return, cravings might resurface, and the fear of regaining weight can feel overwhelming.

But it doesn't have to be that way. With the right strategies, you can build a foundation of habits, mindset, and nutrition that allows you to thrive without the medication.

Why Tapering Off Can Feel Challenging (But Doesn't Have to Be)

Studies have shown that people who stop taking GLP-1 medications often regain most of the weight they lost within six months. This isn't because of a lack of willpower—it's biology.

When you lose weight, your body responds by:

- **Increasing hunger hormones like ghrelin,** making you feel hungrier than before.
- **Decreasing your metabolic rate,** meaning your body burns fewer calories at rest.
- **Triggering survival mechanisms** that encourage your body to regain lost weight.

These biological responses are part of your body's natural defense system against starvation—but in a modern world where food is abundant, they can work against you.

GLP-1 medications help override these signals, but when you stop taking them, your body may revert to its old patterns. This is why it's critical to have a plan in place to manage hunger, stabilize blood sugar, and maintain progress naturally.

The Path to Independence: What You Can Do

Transitioning off GLP-1s is about more than just stopping the medication. It's about creating a new normal—one that's sustainable, empowering, and uniquely yours. Here's how:

1. Rewire Your Relationship with Food

GLP-1s help reduce cravings, but the long-term solution lies in understanding the triggers and patterns that drive your eating behaviors. Emotional eating, stress, and boredom often play a bigger role in weight struggles than hunger itself.

Through mindful eating practices, pattern interrupts, and the strategies you'll learn in this book, you can reshape your habits and take back control.

📖 *For in-depth strategies on breaking emotional eating patterns and managing food triggers, see chapters seven & nine.*

2. Balance Your Biology Naturally

Your DNA Blueprint will show you how your body processes nutrients, which foods fuel you best, and how to stabilize your blood sugar naturally. By aligning your nutrition with your biology, you'll reduce cravings and hunger without relying on medication.

📖 *Learn more about DNA-based nutrition and metabolism in chapters six & 10.*

3. Build a Hunger Management Toolkit

Hunger is natural—it's your body's way of signaling its needs. The key is learning how to respond to hunger in a way that supports your goals. This book will teach you practical tools to manage hunger, from hydration and high-protein meals to strategies for slowing down digestion and staying satisfied longer.

📖 *For practical strategies to regulate appetite and stay in control, see chapters nine & 12.*

4. Strengthen Your Mindset

Your mindset is the foundation of everything. This is where the real transformation happens—not in your body, but in your beliefs. When you shift from relying on external tools (like medication) to trusting in your ability to create change, you unlock a level of freedom that no prescription can provide.

📖 *Mindset is the foundation of success—dive into chapters five, eight, and 13 to reinforce your mental resilience.*

A New Way Forward: Your Journey Beyond GLP-1s

The science of GLP-1 medications is fascinating, and their impact is undeniably powerful. They've likely helped you take some of the most important first steps in reclaiming your health and weight. But the most exciting part of your journey isn't about the medication itself—it's about what comes next.

This is where your story takes center stage.

Imagine yourself six months, a year, or even five years from now. You're no longer relying on a prescription to feel in control of your hunger or cravings. The fear of weight regain is a distant memory. You're living a life that feels free—free from constant hunger, free from dieting, and free from the mental and emotional toll that weight struggles once had on you.

How does that life feel?

For many, GLP-1 medications are a steppingstone—a catalyst for change. But they're not meant to be the destination. They've helped you press the "reset" button on your health, and now, you have the opportunity to take what you've learned, build upon it, and create something extraordinary: a life of vitality, confidence, and true freedom.

The Shift from Quick Fix to Lasting Solution

Let's be honest: The health and weight loss industry are full of quick fixes. We've all seen the promises—"Lose 20 pounds in 20 days!" or "Never feel hungry again!" But you and I both know that quick fixes don't last.

Here's what makes your journey different:

This isn't about a quick fix. This is about a lasting, **holistic solution** that aligns with your body on a cellular level—**your DNA.**

GLP-1 medications have been a powerful tool to quiet the noise, reduce the chaos of cravings, and give you breathing room to focus on yourself. But true transformation doesn't come from a syringe or a pill—it comes from within. It comes from building a life where your health feels sustainable, empowering, and completely under your control.

The DNA Factor: Unlocking Your Unique Blueprint

Here's the truth that most weight loss programs ignore: **Your body is unlike anyone else's.** What works for one person may not work for another, and much of that comes down to your genetics—your DNA Blueprint.

Your DNA is the key to understanding what your body truly needs. It holds the answers to questions like:

- Why do certain foods make you feel energized while others leave you sluggish?
- Why do you store fat in specific areas?
- Why do some workouts leave you feeling strong and empowered, while others drain your energy?

With the insights from your DNA Blueprint, you'll no longer have to rely on guesswork or one-size-fits-all solutions. Instead, you'll have a roadmap that's uniquely tailored to your body's biology. This is the kind of personalized approach that makes long-term success possible—not just for your weight, but for your overall well-being.

For many people, the idea of transitioning off GLP-1 medications can feel daunting. After all, these medications have been your safety net—quieting hunger, stabilizing cravings, and giving you a sense of control that may have felt impossible for years.

But here's what I want you to know: You don't need a safety net when you've built a foundation.

When you shift your focus from relying on medication to creating sustainable habits, you unlock a level of freedom that no prescription can provide. You take the wheel of your own health journey, steering it in the direction that feels aligned with your goals, your values, and your vision for the future.

Your GLP-1 medication journey may have been the beginning, but it's not the destination. You are the driver, and your body has everything it needs to thrive—you just need the right tools to unlock its potential.

Moving Toward Empowerment: Taking the Wheel of Your Health

This book is not just a guide to weaning off GLP-1 medications. It's a guide to creating a life that feels empowering and sustainable. It's about shifting your mindset from *"I need this medication to succeed"* to *"I am capable of sustaining my progress because I trust my body and my plan."*

With the DNAslim approach, you'll discover how to:

- Work with your body, not against it.
- Align your nutrition, movement, and habits with your unique biology.
- Build resilience so you can navigate challenges without relying on quick fixes.
- Focus on holistic health—physical, mental, and emotional.

This isn't just about weight loss. It's about reclaiming your health, your confidence, and your power. It's about stepping into a version of yourself that feels strong, capable, and free.

Your Moment of Choice—Which Path Will You Take?

As you read this, you're standing at a crossroads. One path keeps you where you are—relying on external tools to maintain your progress. The other path invites you to step into the unknown, where you'll build a foundation of habits and strategies that empower you to thrive without medication.

The second path isn't easy, but it's worth it. And you're not alone. This book is here to guide you, step by step, as you navigate this transition. Together, we'll create a roadmap that aligns with your unique needs and goals, so you can move forward with clarity, confidence, and excitement for the future.

Looking Ahead: Rewiring Your Mindset for Sustainable Success

In the next chapter, we'll dive into the most powerful tool for creating lasting change: **your mindset.** You'll learn how to shift from self-doubt to self-belief, how to overcome the limiting thoughts that hold you back, and how to build the mental resilience that will carry you through this transformation.

Let's take the next step together. Your future is waiting.

Your Turn: Integrate & Empower

Mini Exercise: Strengthening Your Mindset and Confidence for Independence

GLP-1 medications can be a helpful tool, but they are not the only solution. True success comes from understanding your body and building habits that allow you to maintain progress long-term. These short exercises will help you strengthen your awareness, confidence, and control over your health journey.

> 📌 *Want to go deeper? A link to the full Whole-BEING Empowerment Workbook is available in the **Resources** section of this book. There, you'll find expanded exercises, action steps, and reflection prompts to personalize your journey. Download your copy to get even more tools and guidance!*

1. Understanding Your Body's Signals

GLP-1s help regulate hunger, but your body also has its own natural hunger and fullness cues. Take a moment to reflect:

- What's one pattern you've noticed about how your body responds to weight loss? (For example, increased hunger, energy crashes, or cravings.)

- 📝 **Quick Action:** For the next day, pay attention to when you feel hunger or cravings. Notice if they are tied to emotions, stress, or routine, rather than true physical hunger.

2. Reframing Fear into Confidence

One of the biggest concerns about tapering off GLP-1s is the fear of regaining weight or losing control. But your progress has never been just about the medication—it's about the habits you've built.

- What is one past success (big or small) that proves you are capable of making and sustaining change?

📝 **Quick Action:** Take a fear you have about this process and reframe it into a statement of confidence. Example:

- Fear: "I'm afraid my hunger will come back too strong."
- Confidence: "I've learned how to nourish my body in a way that stabilizes my appetite and keeps me in control."

3. Strengthening Your Foundation for Independence

GLP-1s may have supported your progress, but now it's time to reinforce your body's natural ability to regulate hunger and metabolism.

- What's one simple habit you can commit to this week that will support your body's natural hunger regulation? (Examples: eating more protein, drinking more water, slowing down while eating.)

📝 **Quick Action:**

- Continue your **15-minute daily walk** to boost energy, regulate blood sugar, and reduce stress-driven cravings.
- Drink more **water**—aim for **1 oz per kg of body weight** (to convert pounds to kg, divide your weight by 2.2). Example: If you weigh 150 lbs. → 150 ÷ 2.2 = 68 kg → 68 oz of water daily.

Final Thought: Taking Ownership of Your Journey

- What is one insight from this chapter that changes how you see your weight-loss journey?

You are not dependent on medication. You are in control of your health. Keep going.

Part 2

Reclaiming Your Mindset and Empowerment

"The stories you tell yourself are more powerful than any craving you've ever had."

— Holli Bradish-Lane

Overcoming Mental Barriers and Becoming Unstoppable

This is the chapter where self-doubt ends—and self-leadership begins. Your mind is the most powerful tool in your transformation. You've likely spent years battling self-doubt, emotional eating, and the fear of failure. Part Two is where we rewire those patterns and rebuild our confidence from the inside out.

We'll dive deep into limiting beliefs, self-sabotage, and the mental roadblocks that have kept you stuck. Through proven strategies like Neuro-Associative Conditioning (NAC) and the 1% Better Principle, you'll learn how to reprogram your habits and emotions around food, movement, and success.

By the time you finish this section, you'll no longer feel like you're fighting yourself. You'll know, with certainty, that you are capable of lasting success—because you'll have the mindset and strategies to back it up.

Let's rewrite your story—starting now.

Chapter 3

The Shift That Changes Everything— Break Free from the Cycle of Self-Doubt

Anita's Story: Why People Stay Stuck

Anita sat in her car, staring through the windshield at the bakery across the street. She wasn't even hungry, but she could already feel the familiar pull—one that had nothing to do with food and everything to do with **comfort, escape, and failure**.

She had tried **so many times** to lose weight. Diet after diet, plan after plan, each one promising this time would be different. Sometimes, she saw progress. A few pounds gone, a few weeks of feeling good... until something threw her off. A stressful day at work. A family crisis. The simple exhaustion of trying so hard, only to feel like she was barely getting anywhere. And each time she slipped, the voice in her head whispered the same cruel truth:

"You're not cut out for this."

She had believed that voice for years. It told her she had no willpower. That her metabolism was broken. That some people were meant to be fit and healthy, but she wasn't one of them.

But the deepest wound wasn't just the weight or the struggle with food. It was the **certainty** that she was **failing at something everyone else seemed to figure out so easily**.

Anita wasn't lazy. She wasn't weak. But she was stuck in **learned helplessness**—a psychological state where past failures had convinced her that success was out of her control.

For years, she believed that if she just **tried harder**, she could break free. But effort alone wasn't enough. What she needed wasn't another diet or exercise plan. What she needed was a **new set of rules**—ones that set her up for success, rather than guaranteeing her failure.

The Run That Proved Her Worst Fear True

A month before that moment in the car, Anita had made a decision.

"I'm going to stop making excuses and just do it. No more waiting. No more easing in. I just have to push myself harder."

She laced up her old sneakers and **went for a run.**

Not a short run. Not jogging around the block. **A six-mile run.**

She made it through the first half mile fueled by determination, but by mile two, her lungs were on fire, her calves burned, and she had to stop. **She felt humiliated. Weak. Angry at herself for thinking she could do this.**

The next morning, her entire body ached. Walking up a single stair was painful—three steps felt like a mountain.

And just like that, the old belief crept back in.

"You're a failure. You're not meant for this."

She drove straight to the doughnut shop. Bought a baker's dozen. Ate seven before she even realized what she was doing.

Sitting in her car, stomach aching, shame settling in, she made a decision—the same one she had made a dozen times before.

"I give up."

That moment, more than anything, reinforced the **lie she had carried for years**—that no matter what she did, she would always end up back here.

What Anita didn't know then was that **her failure wasn't proof that she couldn't succeed. It was proof that she was following the wrong set of rules.**

The Rules You Live by Shape Your Results

Most people believe that **lack of willpower** is what holds them back. That if they could just be more disciplined, push harder, or resist temptation better, they'd finally reach their goals.

But willpower is **exhausting**, and when it runs out, people fall back on **whatever rules they've been living by—whether those rules serve them or not.**

Anita's rules were clear:

- If she couldn't do something **perfectly**, she had failed.
- If she made a mistake, the entire effort was **ruined**.
- If she couldn't lose weight easily, it meant she wasn't **meant to be fit**.

She didn't **choose** these rules. They were built from years of disappointment, self-judgment, and experiences that reinforced the idea that she wasn't capable of success.

But rules are not facts. **Rules can be rewritten.**

The Rules Realignment Exercise: Shifting from Limiting Beliefs to Empowering Standards

What if I told you that you have already succeeded at something that once felt impossible?

Think back to a time when you doubted yourself—but did it anyway. Maybe it was a career move, getting through a tough class, parenting through exhaustion, overcoming an injury, or even learning to drive a car. At first, it probably felt overwhelming. You may have struggled, failed, or wanted to quit. But eventually, you figured it out.

You adjusted.

You got stronger.

And one day, what once seemed impossible became second nature.

That success didn't happen overnight. It happened because you kept showing up.

Now, ask yourself: **What if your health journey is the same?**

Right now, your brain may be telling you that change is too hard. That it won't last. That failure is inevitable. But look at your past. You've already proven that you can adapt, grow, and overcome challenges. This is no different.

Challenge:

Think about one thing you've accomplished in life that once felt out of reach. What did you tell yourself at the beginning? How did you push through? And what did it feel like when you succeeded?

That feeling—the one where you realize you *can* do hard things—is the mindset shift that will carry you through this journey.

Rewriting Internal Rules for Success

Anita's biggest shift didn't come from another diet or workout plan. It came when she changed the rules she lived by.

She stopped telling herself, *"I should exercise more, but I don't have time,"* and replaced it with, *"I choose to move my body in ways that feel good, even if it's just 10 minutes."*

She stopped saying, *"I have no self-control around food,"* and replaced it with, *"I am learning to eat in a way that fuels my body and my energy."*

She stopped believing, *"I'm just not the kind of person who can be fit,"* and replaced it with, *"I am becoming stronger, one small step at a time."*

Anita's new rules made room for success in ways her old rules never did.

And that's the shift *you* need to make, too.

Changing Strategy for Real-World Success

Once Anita had rewritten her subconscious intellective rules, she changed her approach.

Instead of trying to run six miles on day one, she committed to walking for 10 minutes a day.

Instead of beating herself up for eating something "bad," she focused on making her next meal a better choice.

Instead of quitting when she felt discouraged, she reminded herself that progress isn't about perfection—it's about persistence.

She set herself up for success, instead of making success impossible.

And when she inevitably had setbacks, she didn't see them as proof that she was failing—she saw them as proof that she was learning.

Success Is Not Magic—It's Built One Choice at a Time

Anita didn't transform overnight. But once she rewrote her rules and took small, consistent actions, she built confidence, resilience, and sustainable habits.

She stopped seeing weight loss as an impossible goal and started seeing it as a process she was in control of.

And you can do the same.

Your Next Step:

1. Identify one rule you've been living by that holds you back.
2. Rewrite it into a rule that empowers you.
3. Take one small action today that aligns with your new belief.

Your Turn: Integrate & Empower

Mini Exercise: Rewriting Your Internal Rules and Taking Small, Aligned Actions

📌 *Want to go deeper? A link to the full Whole-BEING Empowerment Workbook is available in the Resources section of this book.*

Rewriting the Rules That Hold You Back

The beliefs and internal "rules" you live by shape your results. If those rules have kept you stuck, it's time to rewrite them.

Step 1: Recognize a Time You Proved Yourself Wrong

Think back to a time when you doubted your ability to succeed at something—but did it anyway.

📖 **Journal Prompt:**

- What's one accomplishment in your life that once felt impossible? (A job, parenting, a skill, a personal challenge, etc.)
- How did you push through self-doubt? What helped you keep going?

📌 **Action Step:**

- Write a short reminder to yourself: *"I've done hard things before. I can do this too."* Keep it somewhere visible.

Step 2: Rewrite Your Internal Rules

Your success isn't about willpower—it's about the standards you set for yourself. Let's change them.

📖 **Journal Prompt:**

- What's one belief or "rule" that has held you back? (Example: *If I can't do it perfectly, I've failed.*)
- How can you reframe this into a new rule that empowers you? (Example: *Every step forward counts, even if it's not perfect.*)

📌 **Action Step:**

- Write down your new rule and say it out loud. Repeat it daily.

Step 3: Take One Small Action Toward Change

Success is built on one choice at a time. Today is your chance to take control.

📖 **Journal Prompt:**

- What's one small action you can take today to reinforce your new belief? (Example: If your new rule is *I prioritize building a stronger body,* commit to a 10-minute body weight, resistance training session.)

📌 **Action Step:**

✔ Continue your daily 15-minute walk—movement reinforces your belief in progress.

✔ Keep your new rule visible and repeat it each morning.

✔ Stay hydrated—small, healthy habits add up to momentous changes.

☑ **Final Reflection:**

- What's one key insight from this chapter that shifts how you see yourself?

📌 **Want more support?** Check out the full Whole-BEING Empowerment Workbook in the Resources section for expanded exercises and deeper transformation strategies.

Chapter 4

Shifting Your Mindset for Success— Rewiring Your Thinking to Thrive Beyond GLP-1s

"You have power over your mind—not outside events. Realize this, and you will find strength."

— Marcus Aurelius

Why Mindset Matters More Than You Think

Every great transformation begins in the same place: your mind.

Think about this: How often have you started a weight-loss plan only to abandon it weeks later? If you're like most people, the issue wasn't the plan itself—it was your mindset. The thoughts you think every day shape your actions, and those actions determine your results.

Our brains are wired to seek comfort and avoid hard things (pain). This means that when the process of change feels overwhelming, your mind will naturally resist—even when you desperately want to succeed. The key to overcoming this resistance is rewiring your mindset so that healthier choices feel empowering instead of burdensome.

This isn't about willpower—it's about science. Your brain operates through neural pathways—connections strengthened by repeated thoughts and actions. Every time you think, *"I'm not good enough,"* or *"I always fail,"* those negative beliefs create a stronger pathway in your brain. But here's the good news: you can create **new pathways** by practicing new, empowering thoughts like, *"I'm resilient,"* or *"I'm learning to trust myself."*

For example, instead of thinking, *"I blew it. I'll never get this right,"* you can retrain your mind to say, *"That was one misstep, but I'm proud of myself for continuing to move forward."*

This process of rewiring your thoughts is powerful, and over time, your mindset will become your greatest asset—a force that drives you toward lasting success.

Chris Nikic and The Power of 1% Better Every Day

Let me tell you about someone extraordinary: Chris Nikic.

Chris is an athlete who happens to have Down syndrome, and in 2020, he became the first person with Down syndrome to complete an Ironman triathlon. For those unfamiliar, an Ironman consists of a grueling 2.4-mile swim, a 112-mile bike ride, and a 26.2-mile marathon—all completed within 17 hours. It's a feat that requires an incredible blend of training, physical endurance, mental resilience, and sheer determination.

What's Chris's secret? The philosophy of *"1% Better Every Day."*

Chris didn't focus on achieving the impossible overnight. Instead, he committed to improving by just 1% every day. Whether it was swimming a little farther, biking a little faster, or practicing transitions between events, he embraced small, consistent improvements. Over time, those tiny changes added up to something monumental.

Chris's mindset is a true demonstration of what's possible when you commit to progress over perfection. He didn't let his circumstances define his potential. Instead, he reframed challenges as opportunities to grow stronger.

How can you apply this to your own journey?

- Don't aim for overnight transformation—focus on being 1% better than you were yesterday.
- Celebrate the small wins, like choosing a healthy snack or completing a short walk.
- Remember that progress is cumulative. Each small step builds on the last, leading to extraordinary results over time.

Chris's story reminds us that success isn't about being perfect. It's about showing up, doing the work, and believing that you are capable of more than you think.

Common Mindset Barriers That Hold You Back

Your mindset is the foundation of your health journey, but certain barriers can hold you back. Let's explore the most common ones—and how to overcome them.

1. The "All-or-Nothing" Mentality

Leah, a 42-year-old attorney, used to believe that success required perfection. Every time she "messed up," like indulging in pizza at her son's soccer game, she'd think, *"I've already blown it—what's the point?"* That one slip-up would spiral into a weekend of overindulgence, leaving her frustrated and ashamed.

This "all-or-nothing" mentality is one of the most common mindset traps. But here's the truth: **Progress isn't black and white—it's about consistency over time.**

Imagine you're climbing a staircase. If you stumble on one step, you don't throw yourself to the bottom. You steady yourself and keep climbing. The same applies to your health journey. A single misstep doesn't erase all your progress.

Reframe it: Instead of thinking, *"I've ruined everything,"* say, *"This was one choice, and I can make a better one at my next meal."*

2. Fear of Failure and Regaining Weight

Fear of failure is one of the biggest obstacles in any health journey. It's not just about weight regain—it's about the worry that all your hard work could unravel, that old habits might creep back in, or that without the structure of a plan (or a medication), you'll lose control. It's that nagging voice that asks, *What if I can't do this on my own?* or *What if I slip up and undo everything I've worked for?*

This fear is normal, but it's also based on a **false assumption**—the idea that success is a fragile, all-or-nothing state. In reality, **lasting progress isn't about never slipping up; it's about having the tools and resilience to keep going.**

One of the most powerful ways to overcome this fear is through **understanding your body and working with it instead of against it**. For example, if your genetic profile reveals a predisposition for sugar cravings (linked to the FTO gene), you can proactively structure your meals with protein, fiber, and healthy fats to maintain balance and reduce temptation. When you have knowledge and strategies tailored to your unique biology, the fear of the unknown fades—and confidence takes its place.

Reframe it: Instead of saying, *"I'm afraid I'll fail,"* shift your mindset to, *"I'm learning what works for my body, and I trust myself to keep moving forward."*

3. The Perfection Trap: Redefining Success

Perfectionism has a sneaky way of sabotaging progress, even when it's disguised as a strength. It convinces you that if you can't do something perfectly, it's not worth doing at all. This mindset creates an impossible standard, leading to burnout, frustration, and ultimately, inaction.

Take Valerie, for example. A 36-year-old teacher and single mom, she decided to tackle her health journey with full force. She meticulously planned every meal, exercised six days a week, and logged every calorie. But when her daughter's Saturday afternoon birthday party threw her off schedule, she skipped a workout, ate a slice of cake, and immediately spiraled into guilt. In her mind, she had "failed." By Monday, she'd abandoned her plan altogether.

This is the perfection trap. It tricks you into thinking that success is about flawless execution, rather than consistent effort over time.

But here's the truth: **Progress is messy.** It's not linear. You'll have setbacks, unexpected life events, and days when you simply don't feel like doing the work. And that's okay.

How to Break Free from the Perfection Trap:

1. **Shift Your Focus from Perfection to Consistency:** Instead of aiming to be perfect, aim to show up. Showing up—even imperfectly—is what creates momentum.

Reframe: Instead of saying, "I missed a workout; I've ruined my progress," say, "I missed one workout, but I'll pick back up tomorrow."

2. **Practice the 80/20 Rule:** Give yourself permission to succeed 80% of the time. That leaves room for life's curveballs while keeping you on track.
3. **Celebrate Small Wins:** Did you eat one healthy meal today? Walk around the block? Choose water instead of soda? Those are wins! Acknowledge and celebrate them as proof that you're moving forward.
4. **Adopt a Learning Mindset:** Mistakes aren't failures—they're data. When something doesn't go as planned, ask yourself: *What can I learn from this? How can I do it differently next time?*

Valerie eventually learned to let go of perfection. She realized that eating one slice of cake didn't erase her progress—it was simply part of living a balanced life. By embracing consistency over perfection, she found a rhythm that worked for her, and her progress became sustainable.

4. Emotional Eating and Self-Sabotage: Healing Your Relationship with Food

Emotional eating is one of the most common struggles people face. It often feels like food is the only way to soothe stress, anxiety, boredom, or even loneliness. But here's the truth: emotional eating isn't about food—it's about what food represents.

Let me introduce you to Carlos, a busy IT professional who worked 10-hour days. For Carlos, food was comfort. After a long, exhausting day, he'd collapse on the couch with a pint of ice cream or a bag of chips. It wasn't hunger driving him—it was the need to unwind, to feel in control, and to fill the emotional void created by stress.

Why We Self-Sabotage

Self-sabotage often happens when there's a gap between what you *say* you want (e.g., weight loss, better health) and what you *believe* you deserve. If, deep down, you associate weight loss with discomfort, attention you don't want, or fear of failure, your brain will create excuses to keep you stuck.

DNA Insight: Some genetic markers, like those tied to the **COMT gene**, influence how your body processes stress. If you're genetically predisposed to higher cortisol levels, you might feel the effects of stress more intensely—making you more likely to turn to food as a coping mechanism.

Breaking Free from Emotional Eating

1. **Identify Your Triggers:** Is it stress? Boredom? Sadness? Journaling can help you pinpoint what emotions are driving your eating habits.
2. **Interrupt the Pattern:** The next time you feel the urge to eat emotionally, pause. Take five deep breaths, drink a glass of water, or step outside for fresh air. These small actions can help break the autopilot response.
3. **Find Healthy Alternatives:** Replace emotional eating with activities that truly nourish you. For Carlos, it was playing his guitar or going for a walk after work.
4. **Reframe Your Relationship with Food:** Food is not your enemy, nor is it your therapist. See it for what it is: fuel for your body and an occasional source of joy—not a solution to emotional pain.

Carlos learned to ask himself one simple question before eating: *"Am I hungry, or am I looking for comfort?"* Over time, he discovered healthier ways to cope with stress, and food lost its hold over him.

Anatomy of a Binge and Breaking the Cycle

Picture this: It's Friday, and you've had a long, stressful week. You planned to eat healthily, but someone brought donuts to the office. You tell yourself, *"Just one."* But that one donut leads to another... and another. Before you know it, you've eaten half the box. Guilt sets in, and your inner critic whispers, *"You've already blown it—might as well keep going."* The cycle spirals from there.

Breaking Down the Cycle:

1. A Misstep Happens: You eat something unplanned.
2. You Feel Discouraged, Like You Failed, and Depressed: Guilt and shame trigger emotional eating.

3. **The Cycle Repeats:** Leading to more overeating, reinforcing feelings of failure.

How to Break the Cycle:

1. **Forgive Yourself Immediately:** The first step is to stop the guilt spiral. Tell yourself, *"This is one choice in a lifetime of choices. I can reset right now."*
2. **Pause and Reflect:** Journaling can be a powerful tool. Write down what triggered the binge and how you felt before, during, and after. Awareness is the first step to change.
3. **Refocus on Your Next Action:** Take one small step to reset—drink a glass of water, go for a walk, or prepare a healthy snack. These actions remind your brain that you're still in control.
4. **Reward Yourself for Breaking the Cycle:** Celebrate your victory, even if it's small. Treat yourself to something non-food-related, like watching your favorite show or taking a relaxing bath.

The Power of Reframing Your Perspective

The way you talk to yourself shapes your reality. Reframing is the process of changing the meaning you attach to a situation, and it's one of the most powerful tools for mindset transformation.

Take Chris Nikic's story. Chris doesn't see his Down syndrome as a limitation—he sees it as an opportunity to inspire others. He reframes his challenges into fuel for his success.

Reframing in Action:

1. **Identify the Negative Thought:** For example, *"I'll never lose this weight; it's too hard."*
2. **Challenge Its Truth:** Ask yourself, *"Is this really true? Have I overcome hard things before?"*

3. **Replace It with an Empowering Statement:** Swap the negative thought with one that motivates you: *"I've overcome hard things before, and I can do it again."*

Example Reframes:

- *"I messed up today."* → *"I'm proud of myself for noticing the pattern and taking steps to change."*
- *"Exercise feels like a chore."* → *"I'm grateful my body can move, and I'm choosing to honor it."*

Reframing helps you shift from feeling stuck to feeling empowered. When you change the *way* you look at challenges, those challenges stop being barriers and become steppingstones to your growth.

Looking Ahead

Your mindset is your most powerful ally on this journey. In the next chapter, we'll dive into practical strategies for managing hunger, cravings, and emotional triggers—equipping you with the tools to thrive in every situation. Remember: every thought, every choice, and every step you take is building the foundation for a healthier, more empowered you.

3 Steps to Lasting Change

Reclaiming control over your health and body isn't just about what you eat or how much you exercise—it's about transforming how you think, what you believe, and how you approach your life. These three steps will help you build the foundation for a healthy, empowered life—one that lasts well beyond your time on GLP-1 medications.

Change isn't about doing more—it's about becoming more. It's about stepping into your power, breaking free from old limitations, and creating a life that aligns with the best version of yourself. These three steps—raising your standards, changing your beliefs, and changing your strategy—will act as your guide to lasting transformation. Each step builds on the next, creating a solid foundation for long-term success.

Let's dive deeper into the process.

Step 1: Raise Your Standards

What have you been settling for in your life? Where have you allowed yourself to believe that "good enough" is the best you can do? Real, lasting change begins with a decision: to demand more from yourself and to believe that you are worth the effort.

Here's the truth: Your current standards determine your results. If you're ready for something extraordinary, it starts by setting a higher standard for how you treat yourself, care for your body, and live your life. To create a life of vitality, confidence, and energy, you need to set a higher standard for how you treat yourself and your body. Transformation starts by deciding that the current state of your health and habits is no longer acceptable. It's not just a wish, a dream, or a goal—it becomes a **MUST**.

When you raise your standards, you redefine what you're willing to tolerate. You don't settle for excuses, procrastination, or self-sabotage. Instead, you commit to showing up for yourself—because you know you're worth it.

Rules Realignment: From "Should" to "Must"

Take a moment to reflect on the rules you've been living by. Are they guiding you toward the life you want, or are they holding you back? For instance:

- Do you tell yourself, "I *should* eat healthier," but often break that rule?

- Do you think, "I *shouldn't* skip workouts," but find reasons to skip anyway?

Here's the difference:

- A "should" is optional—it's easy to ignore.
- A "must" is non-negotiable—it's something you live by.

For example:

- **Should:** "I should drink more water."
- **Must:** "I must drink at least eight glasses of water daily to fuel my body."

When you raise your standards, you set the tone for everything else to follow. You stop tolerating behaviors and habits that don't serve you. You start to expect more from yourself, and you'll find that you rise to meet those expectations.

Step 2: Change Your Limiting Beliefs

What's holding you back isn't a lack of willpower, discipline, or even knowledge—it's your beliefs.

What Is a Belief?

A belief is simply a feeling of certainty about what something means. But here's the thing—beliefs are not facts. They are often interpretations of past experiences that you've carried forward. Unfortunately, limiting beliefs—like "I'll always fail at weight loss" or "I'm not good enough"—are often built on shaky foundations, like past missteps or societal messages that made you doubt yourself. But here's the truth: **The past does not equal the future.**

The Fence Post of Limiting Beliefs...

Let me diverge for just a minute to make a point.

Imagine a young horse colt, full of energy and potential, being trained to wear a bridle and accept the new feel of a lead rope. Over time, the colt is tied to a sturdy fence post. At first, it resists, pulling against the rope, testing its strength. But eventually, it learns: the rope and the post mean it cannot go anywhere.

As the colt grows into a powerful horse, its strength and size far surpass the restraint of the fence post. Yet, when tied by the same lead rope and bridle, it stands quietly, no longer resisting. The horse has been conditioned to believe it is restrained, even though it now possesses the power to break free with ease.

This is the power of limiting beliefs. Like the horse, we can become tethered to ideas and patterns that no longer serve us, conditioned by past experiences to think we are held back when, in reality, we have the strength to break free. Recognizing these beliefs is the first step to reclaiming our potential and stepping fully into the freedom and power that's always been ours.

Breaking Free from the Beliefs Holding You Back

To move forward, you must challenge the thoughts that tether you.

- Do you believe your situation is permanent? ("This can never change.")
- Do you see it as pervasive? ("Because of this, everything in my life is ruined.")
- Do you take it personally? ("There's something wrong with me.")

These beliefs are not truths—they're stories you've been telling yourself. It's time to rewrite the narrative.

Replace limiting thoughts with empowering ones:

- "The past does not define me. I'm creating my future."
- "I've succeeded in hard things before, and I can do it again."
- "I am capable, resilient, and strong."

The single most powerful predictor of your success is your belief that you will succeed. Build a foundation of belief, and you'll be unstoppable.

Shift from Wishing to Certainty

Wishing leaves room for doubt. Certainty says, **"This is happening, and I'm making it real."**

When you anchor yourself in the belief that transformation is possible for you, everything changes.

Step 3: Change Your Strategy

Once you've raised your standards and reprogrammed your limiting beliefs, the next step is to change your strategy. This step is about execution—creating a plan that not only aligns with your new mindset but also supports you through life's inevitable challenges.

A great strategy is both sustainable and flexible, allowing you to stay on track without feeling overwhelmed or restricted. Let's break it down.

Planning Sustainable Actions for Challenging Situations

Life doesn't stop just because you're working on a transformation. Social gatherings, unexpected stress, and curveballs will arise—but they don't have to derail your progress. With the right strategies, you can navigate these situations with confidence and stay aligned with your goals.

1. Get-Togethers and Social Events

Social gatherings often revolve around food, making them a potential minefield of triggers. But they're also opportunities to connect and enjoy yourself without losing sight of your progress.

Your Strategy for Success:

- **Plan Ahead:** Before attending an event, think about what and how much you'll eat. If you know there will be indulgent options, decide in advance to enjoy a small portion guilt-free.
- **Eat Before You Go:** Don't show up to the event hungry. Have a healthy snack with protein and fiber beforehand to curb cravings.
- **Bring a Dish:** If possible, contribute with a healthy option you'll feel good about eating.
- **Focus on Connection, Not Food:** Shift your mindset. Instead of seeing the event as centered around food, make it about conversations, laughter, and connection.
- **Have a Plan for "Food Pushers":** If someone insists you try their dish, politely decline by saying, "That looks amazing, but I'm already full. Maybe next time!"

2. Nights Out

Restaurants, movie nights, or evenings on the town can feel like a challenge, but they don't have to derail your progress. The key is to approach these situations with intentionality.

Your Strategy for Success:

- **Study the Menu:** Many restaurants post their menus online. Look ahead and choose a meal that aligns with your goals.
- **Practice Portion Control:** Restaurants often serve oversized portions. Don't feel pressured to finish everything on your plate—stop when you're satisfied, not stuffed.

- **Skip the "Extras":** Decline bread baskets, sugary drinks, or high-calorie appetizers. Instead, savor a dish that feels both satisfying and nourishing.
- **Mindfully Indulge:** If you choose to treat yourself, enjoy it fully. Eat slowly, savor the flavors, and stop when you've had enough.
- **Balance the Day:** If you know you'll be eating out later, focus on lighter, nutrient-dense meals earlier in the day.

3. Crisis Situations

Stress, crises, or unexpected emotional triggers can disrupt even the most well-laid plans. It's natural to feel the urge to turn to food for comfort, but with the right strategies, you can navigate these moments in a way that empowers you.

Your Strategy for Success:

- **Pause and Breathe:** Stress activates your fight-or-flight response, making it harder to think clearly. Take five deep breaths to calm your nervous system and create space to make a thoughtful choice.
- **Redirect Your Focus:** Remind yourself that overeating won't solve the crisis—it will only add to your challenges. Instead, ask yourself, "What can I do right now to feel better that doesn't involve food?"
- **Move Your Body:** Physical movement is one of the fastest ways to release stress and change your state of wellbeing. Whether it's a brisk walk, dancing to your favorite song, or doing a few stretches, movement can help you reset.
- **Reframe the Moment:** Remind yourself, "This is temporary, and I've handled challenges before. I'm stronger than this moment."

Avoiding Triggers: The Impulse Breakers

Triggers are the cues—both external and internal—that lead to unhelpful behaviors like emotional eating. Learning to manage your triggers is essential for lasting success.

Understanding Your Triggers

Triggers come in three main forms:

1. External Triggers: Environmental cues like the smell of popcorn at the movies, a candy display at the checkout line, or a coworker bringing donuts to the office.
2. Mental Triggers: Internal thoughts or beliefs, like "I deserve this treat" or "One bite won't hurt."
3. Physical Triggers: True hunger or physical sensations, like fatigue, dehydration, or low blood sugar, which may be mistaken for cravings.

Impulse Breakers: Your Toolbox for Managing Triggers

When a trigger strikes, these techniques can help you regain control:

1. Drink Water: Often, cravings are mistaken for thirst. Sip a glass of water and wait a few minutes to see if the urge passes.
2. Distract Yourself: Engage in an activity that absorbs your attention, like reading, calling a friend, or working on a hobby.
3. Change Your Environment: Remove yourself from the triggering situation. Step outside, take a walk, or move to a different room.
4. Use Pattern Interrupts: Physically change your mental state by clapping your hands, tapping your toes, or standing up and stretching. These actions disrupt the automatic behavior pattern.

5. Replace the Impulse: Find a healthier alternative that satisfies the same need. For example:
   - If you crave something crunchy, choose raw veggies with hummus.
   - If you crave something sweet, try a piece of fruit or dark chocolate.
6. Affirm Your Commitment: Remind yourself of your "why"—we'll cover more about this shortly. Say out loud, "I am in control. I choose health and vitality."

Positive Physiological Changes

Your physical state profoundly influences your mental state. By making small, intentional changes to your physiology, you can break the cycle of triggers and urges.

Strategies to Leverage Your Physiology:

- Smile: Even if you don't feel like it, the act of smiling actually sends signals to your brain that improve your mood!
- Stand Tall, Set Your Shoulders Back: Adjust your posture to convey confidence and control. This simple shift can immediately change how you feel in the moment.
- Deep Breathing: Breathe deeply into your belly for four seconds, hold for four seconds, and exhale for six seconds. This activates your parasympathetic nervous system, calming your mind and body.
- Just Move: When you feel triggered, stand up, stretch, or take a quick walk. Movement releases endorphins, which naturally reduce cravings and elevate your mood.

Creating a Flexible Plan for Real Life

No strategy is perfect. Life will throw challenges your way, and you'll face moments when sticking to your goals feel harder than ever. The key is to have a plan that's both structured and adaptable.

How to Create Your Flexible Plan:

- **Anchor Your Day:** Start with one non-negotiable habit, like a balanced breakfast, a morning walk, or evening journaling. This anchors your day in consistency.
- **Embrace the 80/20 Rule:** Aim to stay on track 80% of the time. Give yourself grace for the 20%—life happens, and perfection isn't the goal.
- **Revisit and Adjust:** Regularly assess your plan. What's working? What's not? Tweak your approach to stay aligned with your goals.

Model Others for Success

If you're unsure of your ability to succeed, look to others who've been where you are and have achieved the transformation you desire. You don't have to reinvent the wheel—others' experiences can provide a roadmap to your own success.

Ask yourself:

- **What beliefs drive their choices?** Are they rooted in confidence, persistence, or self-compassion?
- **How do they approach challenges?** Do they see setbacks as opportunities to learn and grow?
- **How do they view exercise and nutrition?** Is it a chore, or do they see it as a gift—an opportunity to honor their body and health?

When you model the mindset and habits of those who've succeeded, you're learning from their victories and avoiding their mistakes. Success leaves clues, and by following them, you can shortcut your path to transformation.

Finding Your Community: Strength in Numbers

Transformation doesn't have to be a solo journey. In fact, one of the most powerful tools for long-term success is **community**—a group of like-minded people who share your goals, understand your challenges, and celebrate your wins.

Why does community matter?

- **Accountability:** When you share your goals with others, you're more likely to stay committed. It's one thing to promise yourself you'll make healthy choices, but it's another to promise a friend or group cheering you on.
- **Support:** On tough days, your community can remind you why you started and help you keep going. They'll understand what you're facing because they've been there too.
- **Inspiration:** Seeing others succeed can be a powerful motivator. Their progress shows you what's possible and reinforces that you're not alone on your journey.

How to Find Your Tribe

1. **Join a Group:** Look for local or online communities centered around health, fitness, or personal growth. These could include running clubs, weight-loss support groups, or online forums.
2. **Lean Into Your DNA Community:** If you're working with my DNAslim: Whole-BEING Empowerment program, connect with others who are following a similar personalized health journey. Share tips, celebrate wins, and encourage one another.

3. **Find an Accountability Partner:** This could be a friend, family member, or coworker who shares your aspirations. Check in with each other regularly to stay on track.
4. **Create Your Own Circle:** Invite people in your life to join you in adopting healthier habits. You'd be surprised how many people are also looking for support and connection.

Together, We Go Further

Think about a marathon. While it's technically an individual race, runners often find themselves buoyed by the energy of the crowd, the camaraderie of fellow participants, and the encouragement of supporters along the way. Your health journey is no different.

Transformation is easier—and far more rewarding—when shared. Whether it's swapping recipes, tackling a tough workout together, or simply having someone to talk to when you're struggling, community keeps you moving forward.

So, don't be afraid to reach out, join a group, or build your own circle of support. Together, you'll inspire and empower one another to achieve what once felt impossible.

And remember: you're not just finding a community—you're becoming part of one. By showing up, sharing your story, and offering support, you'll leave a lasting impact on others, just as they will on you.

Envision It to Make It Real

Even if you don't know someone to model, you can use your imagination as a powerful tool.

Take a few moments every day to envision your healthiest, most vibrant self. Picture how it feels to wake up energized, confident, and proud of your choices. Imagine yourself thriving in every area of your life.

When you vividly imagine something, your brain begins to believe it's real. Over time, this vision becomes your reality.

You Have the Power to Thrive

Changing your strategy isn't just about forming habits—it's about designing a system that empowers you to navigate life's challenges with confidence and resilience. With the right tools, techniques, and mindset, you'll be equipped to stay aligned with your goals no matter what comes your way.

Remember Chris Nikic's philosophy of being 1% better every day? Just like Chris tackled an Ironman by breaking it into small, manageable steps, you have the power to take control of your journey, one thoughtful choice at a time. This isn't just about transforming your body—it's about creating a life of vitality, freedom, and purpose.

The best part? This plan is uniquely yours, tailored to your needs and designed to help you thrive. Remember, you're stronger than you think—every step you take brings you closer to the life you deserve. Keep going—you've got this!

Effective Goal Setting: Defining Success on Your Terms

When it comes to transformation, goals are your compass. They provide direction, clarity, and purpose—a tangible way to measure progress as you move toward the life you want. But not all goals are created equal. To achieve lasting change, your goals must reflect what truly matters to you, be realistic enough to keep you motivated, and structured in a way that sets you up for success.

Let me tell you a personal story about how goal setting completely changed my life.

My Turning Point: From Overwhelmed to Empowered

Fifteen years ago, I found myself in a place that felt insurmountable. I was 70 pounds overweight, my knees ached constantly, and the career I loved—a demanding healthcare leadership position—felt like it was draining every ounce of energy I had left. I'd been a runner since I was 15, but at that point, even the thought of lacing up my shoes felt impossible.

Then came the breast cancer scare. It started as a routine mammogram, but when the doctor used words like "alarming" and "prepare for the worst," my world stopped. Waiting for the biopsy results was agony. When the results came back negative, my husband and I were overwhelmed with relief—but the experience left a lasting imprint.

I couldn't keep living the way I had been. The fear was a wake-up call. I realized I needed to make changes—not because I wanted to lose weight or run faster, but because I wanted to live. I wanted to be present for my family, feel strong in my body, and embrace a life full of vitality.

But where do you start when the mountain ahead feels so steep? For me, it began with one small, achievable goal: walking for 10 minutes every evening. That single, tiny step eventually led to another—and another. Over time, those small, intentional actions snowballed into the extraordinary.

Start Small to Go Big: The SMART Way

Looking back, I realize that the key to my success was setting **SMART goals**: Specific, Measurable, Achievable, Relevant, and Time-bound. These weren't lofty, vague resolutions like "lose weight" or "get healthier." They were clear and actionable.

Here's how I applied the SMART framework to my journey—and how you can, too:

1. **Specific:** Instead of saying, "I want to be healthy," I committed to something concrete: "I will walk for 10 minutes every evening after dinner."
2. **Measurable:** I tracked my progress. Some nights, it was only a short loop around the block, but every step counted. Seeing that progress fueled my motivation.
3. **Achievable:** I didn't try to run a marathon right away. That would have been unrealistic and discouraging. My first goal was simple: movement without pain.
4. **Relevant:** My goals aligned with my "why." I wanted to decrease my extra weight and knee pain, have more energy for my family, and reduce my long-term health risks.
5. **Time-bound:** I set a timeline. My first goal was to walk for 10 minutes a day for one month. That felt doable—and it gave me something to aim for without feeling overwhelmed.

Lofty Goals Rarely Work—Small Wins Add Up

The truth is lofty goals can backfire. They're often too overwhelming and disconnected from your current reality. Instead, start small. Your first goal should feel so attainable that it almost seems too easy.

Why? Because small wins build momentum. Every time you achieve a goal—no matter how small—it reinforces your belief that you can succeed. It's like adding fuel to a fire. Before you know it, those small wins create a ripple effect, leading to bigger changes than you ever thought possible.

The Power of 1% Better

When I think about the power of small, consistent goals, I'm reminded again of Chris Nikic. Chris didn't start by training for a full Ironman. He started by committing to being 1% better every day. One more lap in the pool. One extra mile on the bike. One more transition practice.

His philosophy of incremental progress resonates deeply with me. It's a reminder that transformation isn't about massive leaps—it's about consistent, intentional steps forward.

Like Chris, I didn't transform overnight. My 10-minute walks turned into 20-minute runs. My focus on eating mindfully helped me discover foods that truly nourished my body. Those small, manageable goals led to losing 60 pounds, easing my knee pain, and reclaiming my health—all while balancing the demands of a high-stress career.

Your Turn: Define Success on Your Terms

So, what does success look like to you? It might not involve running marathons or climbing mountains. Success could mean feeling confident in your clothes, having the energy to play with your kids, or simply feeling at peace in your body.

To start, ask yourself:

- What small change can I make today that aligns with my goals?
- How can I measure my progress?
- Does this goal feel realistic given my current circumstances?
- Why does this goal matter to me?

Remember: This is your journey. Your goals should reflect what's important to you—not what anyone else thinks you should do.

Keep It Simple, But Start

Effective goal setting isn't about creating a perfect plan—it's about taking the first step. Maybe that step is drinking one more glass of water today.

Maybe it's walking to the end of your driveway. Whatever it is, commit to it.

Because here's the thing: small steps, taken consistently, lead to extraordinary transformation. And just like I learned during my own journey, every small win brings you closer to the life—and the health—you deserve.

So, start where you are. Define your goals. Keep them small, keep them SMART, and keep moving forward.

Looking Ahead

Now that you have broken free from self-sabotaging beliefs, it's time to take it even further. In Chapter 5, we'll introduce Neuro-Associative Conditioning (NAC), a powerful method to rewire your brain for long-term success.

Your Turn: Integrate & Empower

Mini Exercise: Reframing Limiting Beliefs, Setting 1% Better Goals, and Identifying Emotional Eating Triggers

> *Want to go deeper? A link to the full Whole-BEING Empowerment Workbook is available in the Resources section of this book.*

Your mindset is your greatest tool for success. These quick exercises will help you shift from self-doubt to confidence, embrace progress over perfection, and build resilience.

1. Reframe a Limiting Belief

The thoughts you repeat shape your reality. A limiting belief can keep you stuck, while an empowering belief can move you forward.

📖 **Quick Reflection:**

- What's one thought that has held you back in the past? *(Example: "I always fail at weight loss.")*
- How can you reframe this into a positive, empowering belief? *(Example: "I am learning how to nourish my body in a way that supports lasting change.")*

✏️ **Write your new belief here:**

📌 **Action Step:** Every morning, repeat your new belief out loud or write it in a journal. Reinforce this positive shift daily.

2. Set a 1% Better Goal

Progress isn't about massive leaps—it's about small, consistent improvements.

📖 **Quick Action:**

- Choose one **simple** action you can take daily to improve your health by just **1%**. *(Example: drinking an extra glass of water, adding five more minutes of movement, or pausing before emotional eating.)*

✏️ **My 1% better goal for this week:**

📌 **Action Step:** Continue your daily **15-minute walk** to boost energy and reduce stress-driven cravings. Pair it with **staying hydrated**—aim for 1 oz of water per kg of body weight. *(To convert pounds to kg, divide your weight by 2.2.)*

3. Identify Emotional Eating Triggers

Emotional eating is often driven by stress, boredom, or habit—not true hunger. Recognizing your triggers is the first step to breaking the cycle.

📖 **Quick Reflection:**

- Think about the last time you ate when you weren't physically hungry. What triggered it? *(Example: stress, boredom, habit, social pressure.)*
- What's one alternative way you can respond to this trigger in the future? *(Example: take a walk, drink water, call a friend, deep breathing.)*

✏️ **My common trigger:**

_____

✏️ **My new response:**

_____

📌 **Action Step:** Keep a **food & mood log** for the next three days. Write down what you eat and how you feel before and after. Look for patterns—this awareness will help you take control.

4. Create Your SMART Goal

To build momentum, set a goal that is **Specific, Measurable, Achievable, Relevant, and Time-bound (SMART)**.

📖 **Quick Goal Setting:**

- What is one **small but powerful** goal you can commit to for the next 30 days? *(Example: "I will walk for 15 minutes, five days a week.")*

✎ **My SMART goal:**

📌 **Action Step:** Write your SMART goal on a sticky note and place it somewhere visible—your bathroom mirror, fridge, or phone screen. Seeing it daily will reinforce your commitment!

**Progress is built on small, intentional steps.
Keep going—you're stronger than you think!**

Chapter 5

The 7 Steps to Lasting Transformation with Neuro-Associative Conditioning (NAC) (Rewiring Your Brain for Sustainable Change)

"Until you change your thinking, you'll recycle your experiences."

— Unknown

What if the key to lasting transformation isn't just in what you do—but how you think?

For many of us, weight loss and health journeys often start with good intentions but get derailed by old habits, limiting beliefs, and emotional triggers. That is because real, sustainable change doesn't happen by simply forcing yourself to eat less or exercise more—it happens when you rewire your brain to support the behaviors you want to adopt.

This is where **Neuro-Associative Conditioning (NAC)** comes in. NAC is a proven method for rewiring the thought patterns and emotional associations that drive your habits. It allows you to break free from old, disempowering behaviors and replace them with ones that align with your goals, creating a foundation for lifelong success.

In this chapter, we will explore the 7 Steps of NAC—a powerful framework for transforming your mindset, habits, and environment. This method was developed by Tony Robbins, a world-renowned speaker and personal development coach. Building on the principles of Neuro-Linguistic Programming (NLP) and behavioral psychology, Robbins introduced NAC as a practical, results-driven approach for creating lasting transformation. You can explore Robbins' foundational work in his book *Awaken the Giant Within* (1991).

Sandra's Story: From Stress Eating to Total Transformation

Sandra, a past client, used this method to completely redefine her health and reclaim her confidence—a powerful reminder that transformation isn't just possible, but within reach.

At 52, she was a performance improvement facilitator, an expert in creating sustainable change. She spent her career analyzing systems, identifying root causes, and implementing strategies that delivered tangible results. Yet, when it came to her own health, she struggled. Stressful workdays led to poor eating habits, and her favorite evening comfort food—a big bowl of buttered popcorn—became a nightly ritual. Over time, the combination of high stress, low energy, and unhealthy habits took its toll, leaving her feeling stuck, frustrated, and disconnected from her best self.

When Sandra came to me, she felt embarrassed. *"I help people create change for a living,"* she admitted. *"Why can't I do it for myself?"*

Through our work together, we incorporated her **DNA Blueprint**, uncovering valuable genetic insights that helped her personalize her approach to health. Using the **7 Steps of NAC** alongside this information, she transformed her mindset, habits, and relationship with food.

Let's walk through Sandra's journey and see how she rewired her thinking, behaviors, and environment to reclaim her health.

Step 1: Decide What You Really Want—Clarity and Vision

Sandra began by getting crystal clear about what she wanted—remember the SMART goal exercise from chapter four? Previously, her goals had been generic: "I should eat better" or "I want to lose weight." This time, we focused on creating a detailed, emotionally compelling vision of her future self.

She pictured herself thriving—walking into work energized and confident, going for evening strolls with her husband, and feeling proud of the healthy choices she made daily. Her vision included being active with her grandchildren, who often asked her to play outside or join them on weekend hikes.

During this process, Sandra also reflected on what had held her back. One key realization came through her DNA Blueprint, which revealed a predisposition to **gluten sensitivity**. This was a game-changer: Sandra had relied on bread, pasta, and her beloved popcorn as staples, never realizing they could be contributing to her sluggishness and low energy.

Her vision became her anchor, and understanding her body's unique needs gave her a sense of empowerment she'd never had before.

Step 2: Get Leverage—Turn Your "'Shoulds"' into "'Musts"'

Next, we worked on turning Sandra's "shoulds" into "musts." In the past, Sandra often thought, "I should cut back on carbs," or "I should stop eating popcorn late at night." But these "shoulds" felt optional, leaving room for excuses.

Using leverage, Sandra connected massive pain to staying the same. She imagined the long-term consequences of her habits—chronic fatigue, worsening joint pain, and the possibility of serious health issues down the line. She also visualized missing out on active, joyful moments with her grandchildren.

At the same time, Sandra focused on the pleasure of transformation. She pictured waking up with energy, confidently managing her workday, and feeling at peace knowing she was fueling her body with foods that supported her health and vitality.

The insights from her DNA Blueprint added to this sense of urgency. Sandra learned that her genetics also indicated a lower tolerance for refined carbohydrates and a higher protein need. This reframed her eating habits: instead of seeing healthy food as deprivation, she began viewing it as a way to honor her body's unique needs.

The Power of Your "Why"

If you want lasting change, your **why** must be bigger than your excuses. It must be deeper than just "losing weight" or "getting healthy"—because those goals alone will not always motivate you when challenges arise.

For Sandra, her true *why* was not just about fitting into smaller clothes. It was about being fully present and active in her life. She wanted to have the energy to play with her grandchildren, to feel strong and vibrant every day, and to break free from the cycle of emotional eating that had controlled her for years.

When setbacks happened (because they always do), she came back to her *why*:

"I deserve to feel good in my body. I want to be the grandmother who runs and plays, not the one who watches from the sidelines."

Faith and Purpose: Connecting Health to a Bigger Mission

Now it's your turn. Take a moment to ask yourself:

- What is your deeper reason for making this change?
- How will your life be different when you reach your goal?
- What will it cost you—physically, mentally, emotionally—if you do not change?

Your *why* is the anchor that keeps you moving forward when motivation fades. Find it, connect with it, and make it bigger than any obstacle you face.

For many, faith is deeply connected to purpose—why we want to be healthy, strong, and present in our lives. Health is not just about looking a certain way; it's about having the energy to fulfill your calling, serve others, and honor the body you've been given.

If faith plays a role in your journey, reflect on this: *How does caring for your health allow you to better show up for the people and purposes that matter most to you?* When you see your well-being as a way to live out your faith and purpose, it transforms the journey from obligation to something sacred.

Step 3: Interrupt Limiting Thought Patterns

Sandra had a longstanding habit of stress eating, particularly in the evenings. After a long day, she would collapse on the couch, tell herself she "deserved a treat," and reach for her go-to comfort food: that big bowl of buttered popcorn.

To break this cycle, we used pattern interrupts—small actions that disrupt automatic behaviors—and a powerful tool called the Compulsion Blowout Technique.

Pattern Interrupts and the Compulsion Blowout Technique

This method works by making a tempting behavior feel repulsive. Instead of resisting a craving with sheer willpower, you mentally rewire your brain's association with the behavior by exaggerating it to an extreme, unpleasant degree.

For example, when Sandra felt the urge for popcorn, she practiced a simple pattern interruption: standing up, taking ten deep breaths, and sipping a glass of water. Then, she engaged the Compulsion Blowout Technique by visualizing her beloved popcorn in a way that made it unappealing—imagining it drenched in grease, sprinkled with dirt, or crawling with insects. By exaggerating the mental image, her brain no longer registered the popcorn as comforting.

These techniques helped Sandra pause long enough to make more intentional choices. Instead of reaching for the popcorn, she began asking herself, **"What do I really need right now?"** Often, the answer was movement or a stress-relieving activity like stretching or listening to music.

Over time, Sandra's evening routine shifted. She no longer craved popcorn out of habit because her brain no longer associated it with relaxation and comfort.

Step 4: Create New, Empowering Patterns

Breaking the old habits was just one part of the equation—Sandra also needed to create empowering new routines that supported her goals.

Using her DNA Blueprint, Sandra began crafting meals and snacks that aligned with her genetic predispositions. She focused on incorporating lean proteins, healthy fats, and fiber-rich carbohydrates while cutting back on gluten-heavy foods and refined sugars.

Sandra discovered new, satisfying alternatives to her beloved popcorn. She started snacking on roasted chickpeas or air-popped popcorn with olive oil and nutritional yeast—both of which gave her the crunch she loved without the drawbacks of her old habit.

These changes not only supported her health goals but also made her feel in control of her choices.

Step 5: Condition the New Patterns Until They are Consistent

At first, Sandra had to put effort into reinforcing her new habits. But over time, they became second nature.

Each morning, Sandra started her day with empowering affirmations, such as, "I am aligned with my health goals," and "I choose foods that energize and nourish me."

She also tracked her wins, no matter how small—like preparing a protein-packed breakfast or swapping her popcorn for a healthier alternative. These daily victories strengthened her confidence and motivation.

Through repetition, Sandra's new habits became automatic, and she found herself genuinely craving the foods and activities that aligned with her goals.

Step 6: Test Your Transformation in Real Life

Sandra had multiple opportunities to evaluate her new patterns in real-life situations.

At a family barbecue, she confidently skipped the gluten-filled rolls and loaded her plate with grilled chicken, veggies, and a fresh garden salad. She did not feel deprived—she felt empowered by her ability to make choices that honored her body.

One evening, after a particularly stressful day, Sandra felt the familiar urge for buttered popcorn. But instead of caving to the craving, she used her pattern interruption, reached for a healthier snack, and reminded herself of her long-term vision.

These moments proved to Sandra that her new habits were sustainable, even under pressure.

Step 7: Build a Supportive Environment for Success

The decisive step was creating an environment that supported Sandra's transformation. She decluttered her pantry, stocking it with the nutrient-dense foods she now enjoys. She also involved her family in her journey, teaching them about the genetic insights she had uncovered and how they shaped her new approach to health.

Perhaps most importantly, Sandra joined the **DNAslim GenoElite community**, where she connected with like-minded individuals who shared her commitment to health and transformation. This supportive network provided accountability, inspiration, and a reminder that she was not on this journey alone.

Your Turn to Transform...

Lasting transformation in seven steps

Sandra's journey shows the power of combining personalized genetic insights with the 7 Steps of Neuro-Associative Conditioning (NAC), now it's your turn:

1. **Decide What You Really Want:** Create a vivid, emotionally compelling vision for your future.
2. **Get Leverage:** Link massive pain to staying the same and immense pleasure to making a change.
3. **Interrupt Limiting Patterns:** Disrupt old habits by using tools like pattern interrupts and the Compulsion Blowout Technique, a principle derived from Neuro-Linguistic Programming (NLP).
4. **Create Empowering Patterns:** Replace unhelpful behaviors with ones that align with your unique genetics and goals.
5. **Condition New Patterns:** Reinforce your new habits through daily repetition and celebration.
6. **Test Your Transformation:** Prove to yourself that your new habits work in real-life situations.
7. **Build a Supportive Environment:** Surround yourself with people, tools, and systems that keep you inspired and on track.

Sandra didn't just change her diet—she transformed her mindset, rewired her habits, and embraced a healthier, more empowered life. You can do the same.

Start with one step, stay consistent, and trust the process. Your transformation is within reach—one choice at a time.

> *"The rules you live by create the results you live with."*
>
> — Holli Bradish-Lane

Your Turn: Integrate & Empower

Mini Exercise: Defining Your Vision, Breaking Old Habits, and Testing New Strategies in Real Life

📌 **Want to go deeper?** A link to the full **Whole-BEING Empowerment Workbook** is available in the Resources section of this book.

Step 1: Define Your Vision for Change

Transformation starts with clarity. Take a moment to visualize your future self—the healthiest, most empowered version of you.

📖 **Journal Prompt:**

- Imagine yourself six months from now, having successfully transformed your mindset and habits. What does your daily life look like? How do you feel physically, emotionally, and mentally?

📌 **Action Step:**

- Write one powerful statement that defines your commitment to this vision.

Example: "I choose to fuel my body with foods that energize me because I deserve to feel my best every day."

Step 2: Discover Your Deep "Why"

Your **why** is the foundation of lasting change. A surface goal like "losing weight" is not enough—it needs to be something deeply meaningful.

📖 **Journal Prompt:**

- Why is this transformation important to you beyond just a number on the scale?
- How will achieving your goal impact your confidence, energy, relationships, or future?

📌 **Action Step:**

- Write your **why** on a sticky note or in your phone's notes app and place it where you will see it daily (mirror, fridge, phone wallpaper).

Step 3: Turn Your "'Shoulds"' into "'Musts"'

Motivation grows when you connect strong emotions to both staying the same and making a change.

📖 **Journal Prompt:**

- What are three consequences of staying where you are? Now, flip it—what are three incredible benefits of taking action today?

📌 **Action Step:**

- Choose one small habit you will commit to daily that aligns with your vision.

(Example: drinking a glass of water before each meal, taking a 10-minute walk, or prepping a high-protein breakfast.)

Step 4: Breaking Old Habits with Pattern Interrupts

Your brain follows familiar pathways unless you interrupt them.

📖 **Journal Prompt:**

- Identify one habit or craving that no longer serves you. What is a small action you can take the next time it arises? *(Example: If stress eating is a habit, take five deep breaths or drink a glass of water before making a food choice.)*

📌 **Action Step:**

- Practice a **pattern interrupt** today. Catch yourself in the moment, pause, and consciously make a different choice.

Step 5: Build New, Empowering Habits

Sustainable success comes from replacing old habits with healthier ones.

📖 **Journal Prompt:**

- What is one new habit that excites you? How will this habit support your long-term success?

📌 **Action Step:**

- Set a **reminder** to reinforce this habit daily *(on your phone, a sticky note, or through an accountability partner)*.

Step 6: Test Your Transformation in Real Life

New behaviors become natural when tested in real situations.

📖 **Journal Prompt:**

- Think of an upcoming challenge *(a social event, stressful day, or late-night craving)*. How will you manage it differently using what you have learned?

📌 **Action Step:**

- Commit to testing one new response this week and observe how it feels.

Step 7: Strengthen Your Environment for Success

Your surroundings shape your behaviors—set yourself up to win.

📖 **Journal Prompt:**

- What is one change you can make to your environment to support your success? *(Example: Clearing out processed snacks, organizing a workout space, or prepping healthy meals.)*

📌 **Action Step:**

- Make **one small change today** to create an environment that supports your goals.

✅ **Final Reflection & Actions:**

- What is your biggest takeaway from this chapter? How will you apply it to your journey moving forward?

📌 **Daily Actions to Reinforce Your Growth:**

✔ **Continue your daily walk**—add **five more minutes** this week for extra movement and reflection time.

✔ **Keep your "why" visible**—read it every morning as a reminder of your commitment.

✓ **Drink water consistently**—aim to meet your daily hydration goal to stay energized and prevent mindless snacking.

📌 **Want more support?** Check out the full **Whole-BEING Empowerment Workbook** in the **Resources** section for expanded exercises and deeper transformation strategies.

Part 3

Personalizing Your Plan for Lasting Results

Leveraging Science to Work with Your Body, Not Against It

"Know thyself. Heal thyself."

— Adapted Proverb

Ever feel like your body is *fighting* you? That is because most weight loss plans ignore your unique genetics, metabolism, and biology. Part Three is where you take back control—by discovering exactly what works for *you*.

We'll introduce DNA-based health coaching—a revolutionary way to customize your nutrition, exercise, and lifestyle based on your genetic blueprint. You'll learn how your body processes food, builds muscle, and regulates hunger—so you can ditch the trial-and-error approach forever.

This section will help you craft a personalized plan that aligns with your biology—one that feels effortless, makes sense for your lifestyle, and delivers sustainable results.

Get ready to stop guessing and start thriving.

Chapter 6

The DNA Connection: Unlocking Your Body's Blueprint for Sustainable Weight Loss

"Personalization isn't a luxury. It's the foundation of real health."

— Holli Bradish-Lane

Now that you've explored the foundations of lasting transformation—rewiring your mindset, creating sustainable change, and building empowering habits—it's time to take things to the next level. This is where **the science of you** takes center stage.

The Science of You: How DNA Unlocks Your Weight Loss Potential

Imagine for a moment that your body came with an instruction manual—a personalized guide to your unique biology. What if you could uncover the specific blueprint for how your body thrives? This is no longer the stuff of science fiction. Thanks to groundbreaking advancements in genetic research, we are now able to use the power of DNA to unlock the secrets to your optimal health and weight loss journey.

Up until now, much of your approach to health and weight loss has likely been based on trial and error—following popular diets, testing different workout routines, and hoping for the best. But **what if you could remove the guesswork?** What if you had **scientific insight** into exactly what your body needs to burn fat efficiently, regulate hunger, and thrive at a cellular level?

In this chapter, we will explore how your **DNA Blueprint** provides this missing piece. You will learn how a simple, non-invasive test can reveal:

- How your body processes carbohydrates, fats, and proteins.
- Your genetic predisposition for cravings, hunger regulation, and metabolism speed.
- The type of exercise your body responds to best.
- Your sleep patterns, stress response, and even how your genes influence motivation and habit formation.

By understanding these genetic insights, you'll gain a powerful advantage in your weight loss and wellness journey—allowing you to work with your body, not against it.

Let me take you back to where it all began.

The Mapping of Life: The Human Genome Project

In 1990, a group of visionary scientists set out to accomplish what seemed impossible: mapping the entire human genome. Known as the **Human Genome Project**, this ambitious endeavor sought to decode the blueprint of human life—the sequence of over three billion DNA base pairs that define who we are.

Under the leadership of pioneers like Robert Waterson and others, the project united scientists from across the globe with one common goal: to better understand the genetic code that influences everything from our health to how we age. Over the span of 13 years, they achieved what many called the moon landing of modern biology: a full map of the human genome.

What did this mean for the world? It opened doors that were once sealed shut. Scientists could now pinpoint the genetic variations responsible for certain diseases, understand how our bodies respond to medications, and even explore the genetic factors that influence weight, metabolism, and overall health. This monumental discovery launched the age of personalized medicine—a revolutionary shift in how we approach health and wellness.

From the Human Genome Project to Personalized Health Coaching

Fast forward to today, and the impact of the Human Genome Project is no longer limited to laboratories. Its findings are now at the heart of personalized health coaching, transforming how we approach weight loss and wellness.

Your DNA contains the roadmap to understanding how your body works—how it processes food, responds to exercise, and regulates hunger and metabolism. No two people are the same, and no single diet or exercise plan will work for everyone. That is why so many traditional approaches to weight loss fail: they rely on one-size-fits-all solutions, ignoring the unique blueprint inside each of us.

What's even more exciting is that genetic research is constantly evolving, with new discoveries and breakthroughs happening daily. Scientists continue to uncover additional insights into how our genetic makeup impacts every aspect of health, from how we metabolize nutrients to how we manage stress and sleep. These advances empower us to create even more precise and effective strategies for achieving lasting wellness.

Here's the game-changer: By analyzing your DNA, we can uncover the specific genetic factors that impact your ability to lose weight and sustain health. From genetic markers that influence your sensitivity to carbohydrates to predispositions for certain cravings or even food intolerances, this knowledge allows us to create a truly personalized plan—one that aligns with your biology instead of working against it.

Let me put it simply: Your DNA is like a map, and with the right tools, we can use it to chart a direct course to your health goals. Here are just a few examples of how understanding your genetic blueprint can transform your journey:

1. **Customized Nutrition:** Imagine knowing exactly how your body responds to macronutrients like carbohydrates, fats, and proteins. For some people, a low-carb diet may lead to faster weight loss, while others may thrive on balanced carbohydrates but need to focus on healthy fats.
2. **Optimized Fitness:** Your genes can reveal how your body reacts to distinct types of exercise. Are you built for endurance, like long runs or cycling? Or are strength-based activities, like weightlifting, your ideal match?
3. **Managing Cravings and Sensitivities:** DNA can uncover genetic predispositions to cravings, like a heightened preference for sugar, or intolerances to gluten or lactose. These insights can help you make smarter, more satisfying choices.
4. **Sustainable Weight Loss:** By understanding your metabolism and how your body stores and burns fat, we can create a plan that is not only effective but sustainable for the long term.

Real-Life Success Story: Joy's Breakthrough—From Energy Crashes to Thriving

Joy had spent years struggling with stubborn weight gain, unpredictable cravings, and energy crashes that left her feeling drained by mid-afternoon. She had tried everything—low-carb, calorie counting, intense workout programs—but nothing seemed to stick. Frustrated and exhausted, she assumed her metabolism was simply "broken."

Then, she took a DNA test. Her results revealed something she had never considered: her body was genetically predisposed to thrive on **moderate carbohydrates and high protein**—yet she had spent years restricting carbs, which only fueled her cravings and left her feeling deprived. She also learned that her **cortisol response was elevated**, meaning stress played a huge role in her weight retention. Armed with this knowledge, we worked together to build a **personalized nutrition and stress management plan** that fit her genetic profile.

The difference was immediate. Within weeks, Joy felt **more energized, her cravings disappeared, and the weight started coming off—without the exhausting struggle.** By aligning her habits with her body's blueprint, she finally found a sustainable way to eat, move, and thrive. Today, she's no longer chasing diets—she's living in a body that feels strong, capable, and in sync with her natural biology.

"I wish I had done this years ago," Joy told me. "For the first time, I feel like my body is working with me, not against me."

Your transformation can be just as powerful. Let's unlock your unique blueprint and create a strategy that works for *you*—because when you stop guessing and start aligning with your body's needs, everything changes.

Your DNA: The Missing Piece

Traditional diets often fail because they treat weight loss as a one-size-fits-all equation. They do not consider your unique biology, history, or preferences. But your DNA doesn't lie—it gives us the information we need to design a plan tailored to you, ensuring your success is based on science, not guesswork.

The Human Genome Project didn't just map the building blocks of life; it gave us the tools to personalize health and wellbeing in ways we couldn't have imagined. And now, you can use those tools to unlock the healthiest, most vibrant version of yourself.

In the pages ahead, we'll explore how to interpret the blueprint of your DNA and apply it to your weight loss journey. You'll discover why traditional diets do not work for everyone and how you can harness your genetic potential to achieve sustainable results.

The science is here. The tools are ready. And you are about to discover the incredible power of YOU.

A Revolution in Personal Wellness

Imagine being handed a roadmap—a guide designed specifically for your body, detailing the most effective way to eat, move, and live to achieve your healthiest, most vibrant self. That is the promise of DNA-based weight loss. By unlocking your genetic blueprint, you can replace guesswork with precision, frustration with clarity, and wasted effort with sustainable success.

This section is about understanding how your unique biology holds the key to long-term health and vitality. Traditional diets assume that one size fits all, but as science continues to prove, that approach rarely works. The exciting truth is this: Your DNA gives you the power to personalize your health journey in ways that are both effective and empowering.

How Your DNA Impacts Weight Loss

Your genes are like an instruction manual for your body. They don't dictate your destiny, but they do influence how your body functions—how it processes food, burns calories, stores fat, and even responds to exercise.

Here are some of the key genetic factors that impact weight loss:

- **Metabolism:** Some people are genetically predisposed to burn calories more efficiently than others. Others may metabolize certain macronutrients, like carbohydrates or fats, less effectively, which can influence weight gain and energy levels.

- **Hunger and Satiety Signals:** Genes that regulate hunger hormones like ghrelin and leptin affect how quickly you feel full and how often you feel hungry. If you've ever felt like you are "always starving," your DNA may be playing a role.
- **Fat Storage vs. Fat Burning:** Certain genes influence whether your body prefers to store energy as fat or burn it as fuel. Understanding these predispositions allows you to tailor your approach to maximize fat loss.
- **Exercise Response:** Some people's bodies thrive on endurance-based activities like running, while others respond better to strength training. Your DNA can guide you to the type of exercise that suits you best.
- **Cravings and Food Preferences:** Your genetic makeup can even influence your taste buds and cravings, from a sweet tooth to a love for salty snacks.

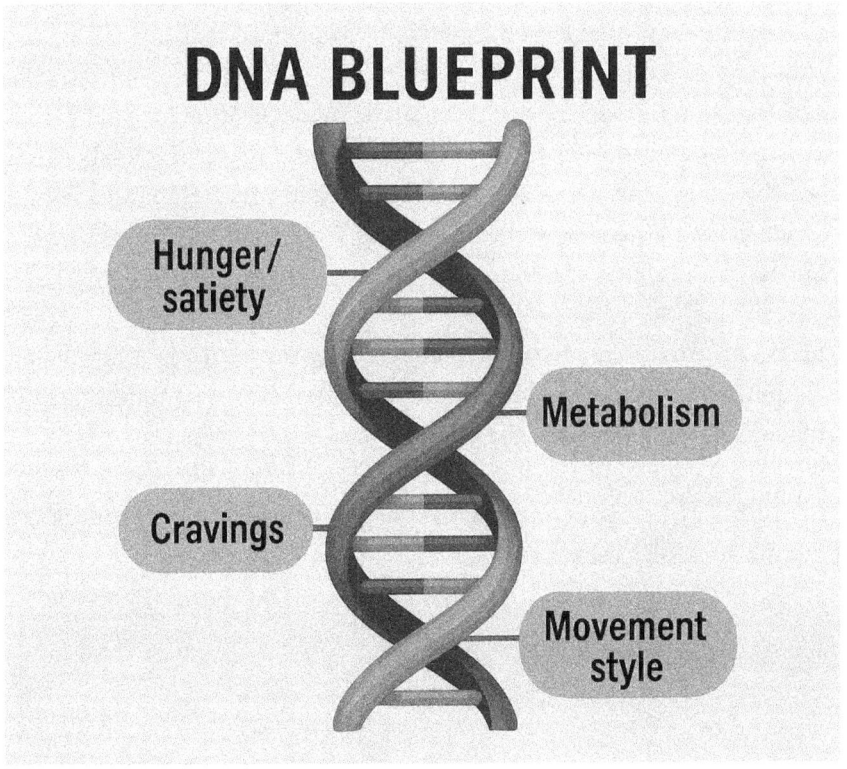

By understanding how these genetic factors work together, we can design a plan that aligns with your body instead of fighting against it.

Why Traditional Diets Fail: The One-Size-Fits-All Myth

Think about all the diets that promise dramatic results—low-carb, low-fat, keto, intermittent fasting. They work well for some people but leave others feeling defeated, frustrated, and stuck in a cycle of yo-yo dieting.

Why? Because traditional diets rely on a one-size-fits-all formula, ignoring the biological uniqueness of each person.

Imagine buying a pair of shoes labeled "one size fits all." Even if they are high quality, they'll pinch your toes or fall off your heels if they don't fit your specific size. Diets are no different. What works for one person may be counterproductive—or even harmful—for someone else.

Take, for example, someone with a genetic predisposition for poor carbohydrate metabolism. Following a carb-heavy diet, even if it's labeled as "healthy," may lead to weight gain, fatigue, and frustration. Meanwhile, someone with a genetic predisposition to store fat may struggle with a high-fat diet like keto or Atkins.

This is why traditional diets fail to deliver long-term success. They do not account for your individual blueprint, and they set you up to feel like the problem lies with you—not the plan.

The truth is, when your approach aligns with your DNA, you unlock the potential for real, lasting change.

Using Your DNA to Optimize Your Nutrition, Fitness, and Long-Term Success

One of the most empowering aspects of DNA-based coaching is that it provides a roadmap tailored to your biology. Here's how:

- **Optimized Nutrition:** If your DNA indicates that you're sensitive to carbohydrates, you can design meals that focus on protein, healthy fats, and fiber while minimizing processed carbs. If you are predisposed to lactose intolerance, you can swap dairy products for alternatives that support digestion and reduce inflammation.
- **Customized Fitness:** Your DNA can reveal whether your body thrives on strength training, endurance exercises, High-Intensity Interval Training (HIIT), or a balance of the three. This ensures that your workouts are effective and enjoyable.
- **Better Recovery:** Genetic insights can identify how your body responds to stress and how quickly you recover from exercise. For someone with slower recovery genes, rest days and active recovery become essential to avoid burnout.
- **Cravings Management:** If you are genetically predisposed to craving sweets or fatty foods, understanding this can help you manage cravings more effectively by balancing blood sugar and choosing foods that keep you satisfied.

With this information, you are no longer relying on trial and error. You're working smarter, not harder.

A Personal Lesson: My Journey from Low-Carb Fatigue to Ironman Success

When I was growing up, I wasn't the naturally athletic kid. I was shy, quiet, and introverted. Sure, I ice skated, played softball, and did the things kids do, but I never thought of myself as someone with "athletic ability." That changed the day, at age 15, a complete stranger challenged and encouraged my best friend and I to lace up our sneakers and run a single mile. It was hard—I mean **really** hard. But something about it stuck. That mile turned into two, and by the time I hit my 20s and 30s, running had become an integral part of who I was.

But as I approached my forties, life started to catch up with me. Between the demands of a high-stress healthcare leadership career and years of putting myself last, I found myself at my heaviest weight—nearly 200 pounds. My knees constantly ached, and I felt like a shadow of the person I used to be. Between the seven inflammation-driven knee surgeries and the cancer scare, I knew something had to change.

I fully committed to making my health a priority, one small step at a time. Slowly but surely, I started seeing results. Over time I lost sixty pounds, my repaired knees stopped hurting, and I felt like I had gained years back on my life.

Through my journey to whole-BEING, I wanted to challenge myself in new ways, so I signed up for a sprint triathlon. The distances were intimidating—a 750-meter swim, 12-mile bike ride, and a 5K run—but I showed up, and I finished. That experience ignited something inside me. Completing that race proved I was capable of so much more than I had believed.

Triathlons became my new passion. As I progressed, I tackled longer and more challenging events, eventually setting my sights on a full Ironman—140.6 miles of swimming, biking, and running. But as I started training for these grueling events, I fell into a common trap: chasing the latest trend.

At the time, the keto diet was making waves in the endurance world. The idea of becoming "fat adapted"—training my body to burn fat instead of carbohydrates—seemed like the key to unlocking more energy for longer races. So, I cut carbs from my diet, embraced the keto lifestyle, and trained harder than ever.

What happened next was one of the most humbling experiences of my life.

Twice, I stood at the start line of a full Ironman race. Twice, I made it through the swim and the bike portion—127 miles of pushing my body to its limits. And twice, I hit a wall halfway through the marathon run. My body shut down. My legs felt like they were encased in concrete, and every step felt impossible. I was forced to quit, just 13 miles from the finish line.

The disappointment was crushing. I had trained, planned, and given it my all, but my body simply could not perform. I questioned myself: Was I not strong enough? Was I asking too much of my body?

Years later, the answers came—not from more training, but from my DNA.

Through genetic testing, I discovered something profound: my body thrives on carbohydrates as its primary energy source. My genetic blueprint revealed that I wasn't well-suited to fat adaptation through a keto diet, especially for endurance events. By cutting out carbs, I had essentially deprived my body of the very fuel it needed to perform. What I thought was an elite strategy turned out to be the exact opposite of what my body needed.

Looking back, it all made sense. During those grueling Ironman events, my body was not failing me—it was trying to tell me something. I just didn't know how to listen.

With this newfound knowledge, I adjusted my nutrition strategy. I reintroduced healthy carbohydrates into my diet and tailored my training plan to align with what my DNA revealed about my body's unique needs. The difference was night and day. I felt stronger, more energized, and capable of pushing my limits without hitting a wall.

This experience taught me an invaluable lesson: no one-size-fits-all approach works for everyone. What works for one person might be the exact thing that derails someone else. When we lean into the science of our own bodies—our unique blueprints—we unlock a path to health and performance that is sustainable, empowering, and freeing.

But more than that, I came to see my body differently. For years, I treated it like something to control, something to push harder, something that would never quite measure up. I had spent so much time frustrated with what my body was not that I had completely lost sight of what it was—something strong, resilient, and worthy of care.

I now see my body as a gift—one that I am responsible for stewarding well. I have come to understand that my health isn't just about personal discipline; it's about honoring the body I've been given. Our bodies are not burdens to fight against or objects to perfect—they are vessels that allow us to live, serve, and fulfill our purpose. Treating my body well, nourishing it, and moving it with intention is my way of respecting that gift.

For years, I ignored what my body was trying to tell me. I pushed through pain, ignored fatigue, and treated rest like a weakness. But I've come to realize that true strength comes from alignment—listening to what my body needs and giving it the care and respect it deserves. My body is not an obstacle; it is a temple. And how I treat it reflects how I show up in my life, my work, and my purpose.

Today, I'm no longer just running or competing in triathlons for the sake of finishing. I'm doing it with confidence, clarity, and the understanding that my body is working with me, not against me. And more importantly, I now see my health as more than just a personal pursuit—it is a way of honoring the life, strength, and purpose I've been entrusted with.

Why My Story Matters

My journey is proof of what is possible when you understand your body on a deeper level. It's not about trying harder—it's about working smarter. Your DNA provides a personalized guide to achieving the health, energy, and vitality you deserve.

You don't have to stumble through trial and error like I did. You can skip the guesswork and go straight to the strategies that work best for your body. When you align your actions with your genetic blueprint, you'll discover that transformation isn't just possible—it's inevitable.

Why Trust This Process?

My passion for this work is deeply personal. As someone who once struggled with weight, sore knees, and the stress of a demanding healthcare leadership role, I know how frustrating it is to feel like you're doing everything "right" and not seeing results.

In my years of healthcare leadership, I watched the field focus on "sick care"—treating symptoms and diseases with medication alone, often after years of damage had already been done. I wanted to flip the script and help people create vitality, strength, and resilience before disease ever had a chance to take hold.

Today, I'm proud to combine my healthcare expertise with cutting-edge genetic science to help others find a path that is sustainable, empowering, and uniquely their own. Through certifications in DNA-based health coaching and years of experience guiding clients through personalized transformations, I have witnessed how understanding your genetics can unlock lasting success.

Adding New Knowledge and Breakthroughs Daily

The field of genetic science is evolving at an incredible pace, with new discoveries and breakthroughs happening every day. What we know today about DNA and weight loss is just the beginning. Each insight adds to our ability to create strategies that are even more precise and effective.

When you choose to work with your DNA, you're not just using today's knowledge—you're tapping into a field of science that's expanding the possibilities for personalized health and wellness.

Getting Your DNA Blueprint: The Process and Next Steps

Now that you understand the power of DNA in optimizing your weight loss and overall health, **you might be wondering: How do I access this information for myself?**

The process is simple, but the insights are profound.

Step 1: A Simple At-Home DNA Test

Your DNA Blueprint starts with a non-invasive at-home test. Using a simple cheek swab, you will collect your DNA sample and send it to a specialized lab for analysis. Within a few weeks, I will create **your personalized genetic report**, which breaks down key insights into your metabolism, cravings, exercise response, sleep, stress, and more.

Step 2: Your Personalized Blueprint Review

Information without guidance can feel overwhelming. That's why your DNA Blueprint isn't about the data—it's about understanding what it means for you.

Once your results are ready, we'll schedule coaching sessions to:

- Walk through your DNA Blueprint and what it reveals about your body's unique needs.
- Identify which nutrition and fitness strategies align best with your genetic strengths.
- Discuss potential challenges (such as a predisposition to insulin resistance or higher cravings) and how to proactively address them.
- Create a personalized action plan for sustainable success.

These one-on-one sessions ensure that your DNA Blueprint becomes an actionable, real-world tool that empowers you—not just an interesting set of numbers.

Step 3: Choose Your Level of Support

Everyone's journey is different, which is why there are multiple ways to apply your DNA insights:

1. DNAslim™ Group Coaching Program – A structured, supportive group coaching experience designed for those who thrive in community. This program takes a deep dive into your DNA Blueprint, walking you through exactly how to apply your genetic insights in a step-by-step format. You will also gain support from like-minded individuals who are navigating the same transition toward sustainable health.

2. Private 1:1 Coaching – For those who prefer personalized attention, private coaching allows for an even deeper level of customization. We will tailor every strategy to your unique genetic profile, lifestyle, and goals, ensuring that your plan fits seamlessly into your daily life.

Whether you choose group coaching or individualized support, the goal remains the same: to help you break free from the cycle of frustration and finally achieve sustainable, lasting results.

Now that you understand how your body's genetic blueprint shapes your health journey, the next step is learning how to take control of your habits, cravings, and neuro-associations. In Chapter 7, we'll dive deep into rewiring the patterns that may have kept you stuck—so you can break free and thrive.

You have come so far. You've done the challenging work—learning, shifting your mindset, and understanding how your body truly works. But knowledge alone isn't enough. **Now it's time to take action.**

Imagine waking up in a body that feels strong, resilient, and energized—not just today, but for years to come. Imagine knowing exactly how to nourish yourself without second-guessing every bite. Imagine the peace of mind that comes from having a personalized plan designed for YOUR body, YOUR goals, and YOUR life.

That's what the DNA Blueprint Coaching Program is about. And now, it's your turn.

Take Control Today: Secure Your Personalized Plan

Spaces in this program are intentionally limited to ensure a highly personalized experience—secure your spot while openings are available.

Each month, I offer a select number of one-on-one strategy sessions for individuals committed to taking control of their health. If you're ready to unlock your unique DNA Blueprint and build a plan aligned with your body's natural strengths, this is your moment.

As a reader of this book, you'll receive priority consideration—but space fills quickly, and availability is first-come, first-served.

Apply for VIP and Individual (1:1) coaching now at *www.ironcrucible-health.com/vip-upgrade-apply-now*

Not quite ready for one-on-one coaching? Join DNAslim group coaching, where we'll work together to apply your DNA insights for sustainable success.

The next chapter of your transformation starts NOW. You've already done the hardest part—deciding you deserve better. Let's build the roadmap to get you there.

You may be thinking... *Is DNA testing really necessary?* Or *What if I don't do the test?*

Questions? Let's meet for a Strategy Session:
https://IronCrucibleHealthScheduling.as.me/

And you know what? Even if you don't take a DNA test, you can still use the principles in this book. The strategies here—balancing blood sugar, managing hunger, and optimizing exercise—apply to everyone. DNA just adds another layer of precision, making it easier to find what works for you faster.

Empowerment and the Freedom of Knowing Your Body

By this point in your journey, you've already discovered the transformative power of mindset, strategy, and sustainable habits. These are the building blocks of lasting change—powerful tools that work whether you're on GLP-1 medications or preparing to step away from them.

But what if you could take this transformation one step further? What if you had a personalized roadmap to guide your nutrition, exercise, and even how you respond to stress? This is where DNA-based insights come in—not as the *only* answer, but as a tool to deepen your understanding of what works for your body.

Even without DNA analysis, the strategies in this book are empowering you to succeed. You've already gained tools to manage hunger, improve your mindset, and handle triggers. But knowing your genetic predispositions adds a layer of precision that allows you to work smarter, not harder.

Take my personal story, for example. I learned that my body thrives on carbohydrates for energy, something I'd unknowingly restricted while trying to become more "fat-adapted" for endurance training. That one insight changed how I fuel my body, not just during races but in everyday life.

For someone else, the key insight might be discovering a predisposition toward gluten sensitivity, which could be contributing to low energy or digestive issues. For another, it might be uncovering a genetic tendency toward slower fat metabolism, guiding them to make subtle yet effective adjustments to their diet.

These insights do not replace the foundation we've built—they enhance it. They allow you to fine-tune your approach, giving you greater confidence in the decisions you make each day.

But remember, even without this information, the tools you've learned so far can create lasting freedom. The frameworks, strategies, and mindset shifts we've covered don't rely on a DNA test—they rely on *you*. With or without genetic insights, you have the power to create a life of health, vitality, and freedom from GLP-1s.

The DNA connection is simply one more powerful way to unlock your body's potential and tailor your efforts. It's about learning to work with your biology in a way that feels intuitive, sustainable, and empowering.

Looking Ahead: Rewiring for True Freedom

As you've seen, unlocking your body's blueprint with DNA insights gives you a powerful edge, but it's only part of the equation. True transformation requires addressing the emotional and psychological patterns that keep you stuck—what we call *neuro-associations*.

In the next chapter, we'll explore how these deeply ingrained associations shape your relationship with food, habits, and even your self-image. You'll learn how to identify the triggers that lead to emotional eating and self-sabotage, as well as actionable strategies to confront them head-on.

Most importantly, you'll discover how to replace disempowering patterns with ones that align with your goals, helping you not just achieve freedom from GLP-1 medications but reclaim your freedom to live life on your terms.

Are you ready to take the next step? Let's rewire your neuro-associations for lasting freedom and strength!

Your Turn: Integrate & Empower

Mini Exercise: Discovering Your Unique Body and Rewriting Your Approach to Weight Loss

📌 *Want to go deeper? A link to the full Whole-BEING Empowerment Workbook is available in the Resources section of this book.*

Your body already holds the answers—you just need to learn how to listen. These exercises will help you start aligning with your unique biology for sustainable, effortless weight loss.

Step 1: Identify Your Personal Weight Loss Struggles

Understanding where you have struggled in the past can help you break free from ineffective approaches.

📖 **Journal Prompt:**

- Think about your past experiences with weight loss. Have you ever followed a diet or workout plan that felt like it *should* work, but didn't? What challenges did you face?
- Did you struggle with cravings, energy crashes, slow metabolism, or feel like your body was working against you?

📌 **Action Step:**

- Write down one frustration you have had in your health journey and reframe it as a learning opportunity. Example: *"I've always struggled with low energy on low-carb diets."* → *"Maybe my body needs a different balance of nutrients."*

Step 2: Your Unique Body—What Feels Right?

Even if you have not done a DNA test yet, your body already gives you clues about what works best for you.

📖 **Journal Prompt:**

- When do you feel your best—energized, strong, and mentally clear? What foods, types of exercise, or daily habits seem to contribute to that feeling?
- When do you feel sluggish, bloated, or drained? Are there patterns in what you eat or do that might be contributing?

📌 **Action Step:**

- Over the next three days, observe how your body responds to different foods and activities. Write down anything that stands out. Do you feel better when you eat more protein? When you get a short walk in after meals?

Step 3: Rewriting Your Approach Based on Science

Now that you understand more about your body's responses, it's time to shift from trial-and-error to strategic action.

📖 **Journal Prompt:**

- What's one way you can stop working against your body and start collaborating with it?
- How would it feel to have a weight-loss strategy designed just for you—one that actually makes sense for your biology?

📌 **Action Step:**

✓ **If you have your DNA results:** Identify one key insight from your DNA Blueprint that you can implement this week.

✓ **If you don't have your DNA results yet:** Choose one small change based on your observations from Step 2. Perhaps evaluate a different macronutrient balance, change your meal timing, or adjust your workout style.

☑ **Final Reflection & Actions**

- What is one "aha moment" you had while thinking about how your body responds to food and exercise?
- What's one small action you will take this week to better align with your body's natural needs?

📌 **Daily Actions to Reinforce Your Growth:**

✓ Continue your daily 15-minute walk—movement enhances metabolism and blood sugar regulation.

✓ Keep a journal of how different foods and habits make you feel—awareness is the first step to personalization.

✓ Stay hydrated—meeting your water needs helps optimize digestion, metabolism, and energy levels.

Chapter 7

Rewiring Your Neuro-Associations for Freedom

How Your Brain Creates Habit Loops and Emotional Triggers

"Discipline isn't punishment. It's radical self-respect."

— Attributed to Glennon Doyle
(popularized via social media and public speaking)

Have you ever felt like you were stuck in a cycle—one where your cravings, habits, or emotions had the upper hand? You've caught yourself reaching for the same comfort foods repeatedly, even when you weren't hungry. Or you've been lured in by that "limited time" sale, unable to resist the urge to act, only to feel regret afterward. These are perfect examples of how your brain creates powerful neuro-associations—connections between emotions, behaviors, and outcomes that shape your choices every day.

The good news? You have the power to rewire these patterns.

This chapter is about breaking free. It's about uncovering why you get stuck, rewiring your habits, and transforming the way you respond to triggers so you can regain control—not just of your eating, but of your mindset and your life.

Let's dig in by exploring a powerful concept: neuro-associations.

Breaking Free from Emotional Eating and Self-Sabotage

Your brain is like a master storyteller. Every experience you have, every emotion you feel, and every decision you make creates a connection—or neuro-association—in your brain. These associations are powerful because they help your brain categorize the world and make quick decisions.

For example, imagine biting into a warm chocolate chip cookie.

You feel the buttery sweetness melt on your tongue, and for a moment, everything feels better.

Your brain lights up with pleasure and comfort.

That experience creates a positive neuro-association with cookies, tying them to happiness or relief from stress.

Now, flip the script. What happens if you eat too many cookies, feel overstuffed, and later regret it? A conflicting neuro-association forms—one where cookies are both a source of joy and frustration. These "mixed emotions" create confusion in your brain, making it hard to break free from patterns like overeating.

Breaking free starts with awareness. Let's look at how scarcity and fear of missing out (FOMO) play a role in these patterns.

Scarcity and FOMO: The Psychological Tricks That Keep You Stuck

Imagine this: It's Girl Scout cookie season. Those thin mints and Samoas you love only come around once a year. You tell yourself, "If I don't buy them now, I'll have to wait another whole year!" So, you stock up—not just one box, but five.

Or consider the email you just got from Kohl's as a VIP customer: "Congratulations! You've been selected to receive 40% off for the next 24 hours!" You suddenly feel special, like you're part of an elite group. You start filling your cart because you don't want to miss this "once-in-a-lifetime" chance.

Do these sound familiar? These are classic examples of scarcity and FOMO at work. They make you feel like you must act immediately or risk losing something valuable. But here is the truth: scarcity is often an illusion.

The Truth About Scarcity

When it comes to food, there's rarely a true shortage. That red licorice in the cabinet or that slice of raspberry pie in the fridge will still be there tomorrow. You don't need to act on the impulse right now.

This is where the *Tomorrow Technique* comes in—a simple yet powerful way to manage cravings.

The Tomorrow Technique: Intelligent Procrastination for Cravings

The next time you feel the pull of a craving, pause and ask yourself:

- Do I really need this right now, or can it wait until tomorrow?
- Will this food still be here later if I decide I want it?

Tell yourself, "I'll enjoy it tomorrow if I still want it." Then shift your focus to something else—a quick walk, a glass of water, or a phone call with a friend.

Most of the time, that craving will fade, and you'll realize you didn't really want the food after all. And if you still want it tomorrow? You can enjoy it mindfully, knowing you made the decision from a place of control, not impulse.

The *Tomorrow Technique* is about giving yourself the space to choose—because when you take control of your choices, you take back your power.

How to Rewire Negative Patterns (and Avoid Self-Sabotage)

"She remembered who she was, and the game changed."

— Lalah Delia

Negative patterns are the automatic behaviors and habits that keep us stuck in cycles of frustration, guilt, and shame. These patterns often show up as self-sabotage—when your actions don't align with your goals, even though you desperately want to succeed.

Think about the last time you overindulged, skipped a workout, or ignored your healthy habits. What was happening in that moment? Were you stressed, bored, or overwhelmed? Did you feel lonely or frustrated? Were you sitting on the couch at the end of a long day, or standing in the kitchen late at night, staring into the fridge because you couldn't sleep?

Negative patterns are not signs of failure—they're conditioned responses that your brain has learned over time. And the good news? What has been learned can be *unlearned*.

John's Story: Late-Night Snacking and Rewiring the Brain for Success

John, a 38-year-old father of two and a successful project manager, came to me feeling defeated. Despite his career success, he struggled with his weight, lacked energy, and couldn't seem to break the cycle of late-night snacking.

John described his evenings as a time of quiet chaos. After putting the kids to bed and catching up on work emails, he would find himself alone in the kitchen, mindlessly reaching for snacks—chips, cookies, or whatever was in the pantry.

"I'm not even hungry," John confessed, "but it's like I can't stop myself. I'll tell myself, 'Just one,' but before I know it, the bag is empty, and I feel horrible. Why do I keep doing this to myself?"

John's story is one many of us can relate to. His late-night snacking wasn't about willpower; it was about unexamined patterns. For John, the trigger was a mix of stress and fatigue. Food became his way of unwinding—a momentary comfort that provided a feel good, dopamine hit but left him feeling worse in the long run.

Through our work together, John began to recognize his negative patterns and, more importantly, rewire them. Here's how we did it...

Steps to Rewire Negative Patterns

1. Identify the Trigger

Rewiring starts with awareness. John learned to pause and ask himself, "What's really going on here?" He realized his snacking was not about hunger—it was about decompressing from the stress of the day.

Common triggers include:

- **Emotions:** Stress, boredom, loneliness, frustration, or even excitement.

- **Situations:** Being alone at night, watching TV, or finishing a stressful workday.
- **Thoughts:** "I deserve this," or "It's been a long day, so why not?"

By identifying the trigger, John could see the root cause of his behavior, rather than just blaming himself for giving in.

2. Disrupt the Pattern

Once you know the trigger, the next step is to disrupt the automatic response. For John, this meant breaking the cycle of walking into the kitchen and reaching for snacks.

We worked together to create simple, actionable tools he could use in the moment, such as:

- **The Tomorrow Technique:** When John felt the urge to snack, he told himself, "I'll have it tomorrow if I still want it." This gave him space to pause and rethink his decision.
- **Physical Movement:** Instead of heading to the pantry, John would take the dog out for a quick walk, stretch, or do 10 push-ups to shift his energy.
- **Change of Scenery:** John started reading in the living room or spending five minutes journaling in a different part of the house to break the kitchen-as-comfort association.

3. Replace the Behavior

Breaking a habit is only half the battle. The key to lasting change is replacing the old behavior with something that aligns with your goals.

For John, we discovered that what he really needed in those moments wasn't food—it was relaxation and a sense of reward. Together, we brainstormed alternative ways he could unwind, like:

- Drinking a cup of herbal tea.
- Journaling about the wins from his day.
- Practicing deep breathing or meditation for five minutes.
- Watching a short, funny video to lift his mood.

These small actions gave John the same comfort he'd been seeking in food, but without the guilt or regret.

The Power of Emotional Conditioning: Breaking Food Attachments

Sometimes, breaking the cycle of a negative pattern requires creating an intense emotional response to the old behavior. By consciously associating the habit with discomfort or consequences, you can rewire your brain to think twice before indulging.

Let me share a personal story that illustrates the power of emotional associations.

When I was in my mid-teens, a few neighborhood friends and I got hold of a bottle of Wild Turkey whiskey one frigid, wintery, New Year's Eve. Feeling bold and sneaky, we mixed it with Coke over ice, laughing as we toasted to our rebellion. At first, it was fun—giddy, tipsy, and carefree. But as the night went on, that fun quickly soured. My stomach began to churn, and before long, I found myself doubled over, vomiting every last bit of comfort I had.

To this day, the smell of bourbon or whiskey evokes an immediate, visceral reaction in me. It's as if I'm transported back to that moment, and the thought of drinking it again is utterly repulsive. That single experience created a permanent association—one that I do not need to remind myself of or reinforce.

Similarly, John began to build his own emotional associations by visualizing the sluggish, regretful feeling he often had after a binge. He practiced pausing and imagining how proud and energized he felt when he made choices aligned with his goals. While you don't need to create a moment of extreme physical discomfort like I did with Wild Turkey, you can use intentional reflection to reframe your patterns.

For instance, the next time you find yourself reaching for the cookie jar, take a deep breath and picture how you will feel afterward. Will it bring the comfort you're seeking, or will it leave you feeling further away from your goals? Then imagine an alternative action—perhaps sipping herbal tea or journaling—giving you the same sense of reward, but without regret.

This practice rewires your brain to associate the negative habit with discomfort and the new behavior with empowerment. Just like my aversion to Wild Turkey, these new associations become second nature, making it easier to align your actions with the vibrant, healthy life you're creating.

The Truth About Self-Sabotage

Here's the most important truth to understand: **Self-sabotage isn't about willpower or discipline. It is about the stories you tell yourself and the patterns you've conditioned over time.**

For John, the story was, "I need snacks to relax." But once he rewired that belief, he realized he was capable of relaxing and rewarding himself in ways that honored his goals.

Your Turn: Break Free from Negative Patterns

Are you ready to rewire your own patterns? Here's how to start:

1. **Identify Your Trigger:** What emotion, situation, or thought is driving your behavior?

2. **Disrupt the Pattern:** Use a tool like the Tomorrow Technique, a change of scenery, or physical movement to break the cycle.
3. **Replace the Behavior:** Choose an action that truly meets your needs—something that aligns with your goals and makes you feel proud.

Remember, this is not about perfection. It's about progress. Like John, you have the power to break free from negative patterns and replace them with choices that support your success.

With every trigger you confront, every pattern you disrupt, and every new behavior you choose, you're taking a step closer to the freedom and vitality you deserve. Keep going—you're rewriting your story one moment at a time!

The Scramble Technique: Creating Negative Food Associations

Another powerful way to rewire your brain is by using the Scramble Technique.

Here's how it works:

1. Think about a favored food you crave but want to avoid, like buttered popcorn or sugary snacks. Close your eyes and vividly imagine it—but this time, distort it. Picture the popcorn dripping with green slime, each kernel swollen and oozing, the artificial butter scent now sickly sweet with a rancid undertone. As you bring a handful to your mouth, the texture feels off—too wet, too sticky. You glance down and realize the slime isn't just coating the popcorn; it's seeping between your fingers, thick and stringy like something decaying. Then the smell hits—sour, putrid, like spoiled milk left out in the sun. The more you try to shake it off, the more it clings, an inescapable, rotting mess that was once your favorite snack.

Or perhaps red licorice is your vice. Imagine your toddler has been swishing it around in the toilet bowl—not just once, but repeatedly, dunking it in and out of the murky water. Now picture the water: cloudy, flecked with bits of toilet paper, carrying a faint, musty odor. Your toddler giggles, squeezing the licorice between their chubby fingers, letting it soak and soften. Then, with a delighted grin, they hold it up to you—wet, slimy, and glistening with a questionable sheen—offering you the first bite.

2. Add more sounds or feelings to the imagery to further repulse you. Imagine the crunch of popcorn turning into the sound of a snapping bug, or the taste feeling gritty and unpleasant. And the licorice... imagine it slick and rubbery, dripping with murky toilet water, carrying the faint scent of mildew and something disturbingly sour as it stretches between your fingers like a soggy earth worm.
3. Continue to intensify these negative associations until the craving fades.

By using this technique, you create new neural pathways in your brain that weaken the pull of those old cravings.

Rewriting the Stories You Tell Yourself About Food and Habits

Imagine you're at a party, and the dessert table is shouting your name. Normally, you'd give in without thinking. But today, you are going to **confront and conquer** that trigger.

1. **Pause and Acknowledge:** Take a deep breath and recognize what is happening. Say to yourself, "I see this trigger, and I know I have the strength to face it."
2. **Visualize Your Victory:** Picture yourself confidently walking past the table, enjoying the party without needing to eat the dessert.

3. **Take Action:** Choose a different focus—strike up a conversation, sip on a sparkling water, or move outside for a few minutes of fresh air.

The more you practice confronting and conquering your triggers, the stronger you will become. Over time, those triggers will lose their power, and you'll gain a sense of freedom you never thought possible.

FREEDOM TOOLKIT
URGE MANAGEMENT TOOLS

PATTERN INTERRUPTS
Break the loop by doing something unexpected to short-circuit the craving cycle.

"Surprise the system—disrupt the urge before it finishes its story."

Examples: clap loudly, sing a silly song, switch hands, snap a rubber band, flip your inner script

FREEDOM TOOLKIT
URGE MANAGEMENT TOOLS

URGE CRUNCHERS
Fast-acting physical or mental actions that redirect your urge before it takes over.

"Change your state, and you change the urge."

Examples: deep breathing, cold water, standing up, brisk walk, body walk, grounding object.

FREEDOM TOOLKIT
URGE MANAGEMENT TOOLS

THE TOMORROW TECHNIQUE
Use intelligent procrastination— delay the action, not deny it.

"Tell your brain: not now, maybe tomorrow."

This softens resistance and gives space for the urge to pass without shame or rebellion.

FREEDOM TOOLKIT:
URGE MANAGEMENT TOOL

THE SCRAMBLE TECHNIQUE
Rewire old associations by mentally scrambling the food craving.

Imagine the food melting, turning gray, covered in dirt, being dipped in vinegar—until it loses emotional charge.

Your Turn: Take Back Control

This chapter is your roadmap to breaking free from the habits and patterns that have held you back. Whether it is scarcity thinking, emotional eating, or persistent cravings, you now have the tools to rewire your brain and reclaim your freedom.

In the next chapter, we will take this transformation even further by diving into **the power of belief and certainty**—and how your mindset can become your greatest asset in building the body and life you deserve.

It's time to take charge of your neuro-associations and rewrite the story you tell yourself. You've got this!

Your Turn: Integrate & Empower

Mini Exercise: Recognizing Triggers, Rewiring Cravings and Conquering Negative Patterns

Want to go deeper? A link to the full Whole-BEING Empowerment Workbook is available in the Resources section of this book.

Step 1: Identify Your Stuck Patterns

To break free from negative habits, you must first recognize what is keeping you stuck.

Journal Prompt:

- Think about a habit you've struggled with—mindless snacking, skipping workouts, emotional eating, or another pattern that doesn't serve you.
- What typically triggers this behavior? Is it stress, boredom, exhaustion, or something else?
- What do you usually tell yourself in those moments? (Example: "I deserve this," or "I'll start fresh tomorrow.")

📌 **Action Step:**

- Write down one trigger you will be mindful of this week. Simply becoming aware of it is the first step to change.

Step 2: Reframe the Scarcity Mindset

Many habits are driven by the false belief that you *have* to act now—or miss out. But when you pause, you often realize that is not true.

📖 **Journal Prompt:**

- Recall a time you gave in to a craving or impulse because you feared missing out.
- Looking back, was that urgency real, or was it just an emotional response?
- How would you manage a comparable situation differently now?

📌 **Action Step:**

- Practice the *Tomorrow Technique* this week. When a craving hits, tell yourself: *If I still want it tomorrow, I can have it then.* Notice how many cravings fade when you give them time.

Step 3: Disrupt and Replace Your Negative Patterns

Breaking a habit isn't just about stopping the behavior—it's about replacing it with something that meets your needs in a healthier way.

📖 **Journal Prompt:**

- Choose a habit you want to change. What small action can you take *instead* of your usual response?

- Example: Instead of snacking out of boredom, try journaling, stretching, or drinking a glass of water.

- How can you make this new habit easy and accessible? (Example: Keep a journal near the kitchen or a bottle of water on your desk.)

📌 **Action Step:**

- The next time you feel the urge to engage in a negative habit, **pause**. Take a deep breath, acknowledge the trigger, and choose a new response that aligns with your goals.

Step 4: Use the Scramble Technique to Weaken Cravings

You can train your brain to lose its attachment to certain foods by changing how you visualize them.

📖 **Journal Prompt:**

- Pick a food you struggle to resist. How can you make it *unappealing* in your mind?
- Imagine it covered in something gross (spoiled milk, mold, dirt). Get as detailed as possible.
- How do you feel about that food now?

📌 **Action Step:**

- Use this *Scramble Technique* when a craving hits. The more you practice, the less appealing that food will become over time.

✅ **Final Reflection: Strengthening Your New Patterns**

📖 **Journal Prompt:**

- What is one key takeaway from this chapter that you'll apply to your daily life?
- How will changing your neuro-associations impact your long-term success?

📌 **Daily Actions to Reinforce Your Growth:**

✓ Continue your **daily movement**—exercise is one of the best ways to disrupt negative habits.

✓ Use the **Tomorrow Technique** to challenge your impulses.

✓ Keep a **visual reminder** of your new habit—place a sticky note on your fridge, mirror, or phone background.

Chapter 8

The Power of Belief and Certainty—Trusting Yourself to Achieve What You Truly Want

Belief is the quiet force that shapes your reality. It is the foundation of every success, every change, and every breakthrough you will ever experience. Without belief, even the best plan will falter. With it, you can defy odds, rise above challenges, and create a life that aligns with your highest aspirations.

For those who've relied on GLP-1 medications, this chapter is a reminder: the power to succeed doesn't lie in a prescription—it lies in you. The medication may have been a catalyst for change, helping you lose weight or regain control of cravings, but lasting transformation is about more than the number on the scale. It is about the belief that you are capable, worthy, and strong enough to carry your progress forward—without relying on external tools forever.

Here, we will dive deep into the transformative power of belief and certainty. Together, we'll explore how to break free from self-doubt, foster unshakable confidence, and use proven techniques like visualization and intentional language to align your mind and body with your goals.

Why Your Belief in Your Success is the #1 Predictor of Results

Belief is the spark that ignites transformation. Without it, even the best tools, plans, or medications will fall flat. But when you genuinely believe in your ability to succeed, everything changes—your mindset, your actions, and ultimately, your results.

The Science of the Placebo Effect and Its Link to Weight Loss

Consider this: Studies on the placebo effect reveal just how powerful belief can be. When people believe a sugar pill will heal them, their bodies respond as if they have received actual medication. This isn't magic—it's the mind signaling the body to align with the belief that healing is happening.

Now imagine channeling that same power into your weight-loss journey. When you deeply believe you are capable of maintaining a light, healthy body, your actions will naturally align with that belief. You'll approach your choices with confidence, overcome setbacks with resilience, and stay committed to your goals, even when progress feels slow.

But here is the challenge: when belief is missing, everything becomes harder. You second-guess yourself, hesitate to take action, and fall back into old habits because deep down, you're not convinced change is possible for you.

For those who have relied on GLP-1 medications, this uncertainty can feel particularly heavy. You might wonder, *Will I be able to maintain my progress without them? What if I gain the weight back?* Let me reassure you: these doubts are normal. But here's the truth you need to hold on to—GLP-1s didn't create your success. They helped quiet the noise of cravings, but every healthy choice you made came from *you*.

It's proof of your strength, not the medication's power. It is a reminder that you are in control of your health. And you have everything it takes to sustain your success, no matter where you are on your journey.

How Your Thoughts Physically Shape Your Biology (Epigenetics & Neuroplasticity)

Belief doesn't just shape your thoughts—it impacts your biology at a cellular level. As Dr. Francisco Torres explains in *Epigenetics and the Psychology of Weight Loss* (2022), "the information we put into our bodies in the form of food, activity, and stressful or enriching stimuli determines the expression that our cells put out."

Think about that: every bite of food, every step you take, every positive or negative thought—it all sends information to your body, influencing how your genes express themselves. This is the science of epigenetics, the bridge between your mindset, habits, and DNA.

But your brain is just as adaptable as your genes. Thanks to neuroplasticity—**the brain's ability to reorganize and form new neural connections throughout life**—you can rewire your habits and cravings over time. When you repeatedly make choices that support your health, such as opting for whole foods over processed ones, your brain strengthens those pathways, making healthier decisions feel more natural and automatic.

Belief plays a vital role here because it is the foundation for the choices you make. When you believe in your ability to create a healthier life, you naturally gravitate toward actions that nourish your body—choosing foods that energize you, engaging in activities that strengthen you, and surrounding yourself with enriching environments.

What's even more exciting is that every single person—no matter their starting point—has the ability to enjoy drastically improved health and well-being by harnessing the power of their choices. Your genes may influence your starting line, but your actions determine the path forward.

Your Unique Potential for Success

This chapter is about helping you foster the belief that *you are capable of sustaining your success*, regardless of your circumstances or past experiences.

Take a moment to reflect on how far you have already come. Perhaps GLP-1 medications helped you take the first step—but you've been the one showing up every day, making the decisions that align with your goals. Those choices are proof of your ability to thrive.

Now, imagine what's possible when you pair that belief with the tools in this book—tools to manage triggers, rewire negative patterns, and personalize your plan for long-term success. You're not just capable of maintaining your progress; you can create a life that exceeds your wildest expectations.

This is your moment to own your power, trust your potential, and believe in the incredible capacity of your body and mind to transform. Let's continue building that belief—one thought, one choice, and one action at a time.

Escaping Learned Helplessness and Reclaiming Confidence

Remember the concept of Learned Helplessness?

Learned helplessness occurs when repeated setbacks convince you that your efforts do not matter. Over time, you stop trying—not because you lack ability, but because you've been conditioned to believe success is out of reach.

Let me illustrate this with a different example. Imagine a child learning to ride a bike. At first, they are wobbly, falling over, and maybe even skinning their knees a few times. If every time they fall, someone tells them, "You're just not good at this—you'll never figure it out," they may start to believe it. Even if they grow stronger and more capable, they might hesitate to get back on the bike, convinced they're bound to fail.

Humans often carry this same hesitance into their weight-loss journeys. If you've faced repeated struggles—diet plans that didn't work, progress that felt temporary, or habits that seemed impossible to break—it's easy to feel like success is unattainable. But here is the truth: *You are not trapped.*

The barriers that once held you back—whether they were cravings, unhealthy habits, or emotional triggers—are not permanent. With the tools in this book, combined with a belief in your ability to succeed, you have everything you need to overcome them.

Angela's Story: Breaking the Cycle of Yo-Yo Dieting

Meet Angela, a 48-year-old case manager and mother of three who had struggled with her weight since her early thirties. Angela felt defeated after years of yo-yo dieting. She had tried countless weight-loss programs, from fad diets to intense exercise regimens, but nothing ever seemed to stick.

"I feel like I'm broken," she confessed during our first session. "I always start strong, but I end up back where I started—or worse. Maybe I'm just not meant to be healthy."

Angela's words revealed a classic sense of *learned helplessness.* Her past experiences had conditioned her to believe that no matter what she did, she couldn't succeed. But together, we worked to change that narrative.

We started small. Angela committed to one new habit: drinking a glass of water before each meal. It wasn't about overhauling her entire lifestyle overnight—it was about showing her that she *could* succeed at something manageable.

As Angela began stacking small wins—like taking short walks in the evenings or choosing a balanced breakfast—her confidence grew. "For the first time, I'm starting to feel like I can do this," she said after just a few weeks. She realized that her previous failures were not a reflection of her abilities—they were a result of unsustainable approaches that didn't honor her needs.

Cultivating Confidence

Building confidence doesn't happen overnight, but it is entirely within your reach. Here's how to begin:

1. **Celebrate Small Wins:** Confidence grows through action. Each healthy choice, no matter how small, is evidence that you are capable. Whether it's drinking more water, walking for 10 minutes, or resisting an unhealthy snack, take a moment to acknowledge and celebrate your efforts.
2. **Focus on Growth, Not Perfection:** Confidence isn't about never failing—it's about knowing you can learn, adapt, and keep moving forward. Angela didn't eliminate all her unhealthy habits at once. She focused on making progress, not being perfect, and that mindset became her greatest asset.
3. **Rewrite Your Internal Dialogue:** The way you talk to yourself matters. Replace self-defeating thoughts with empowering affirmations like:
   - "I am strong, and I am capable of change."
   - "My efforts matter, and I am worthy of success."
   - "I've overcome challenges before, and I can do it again."

Your Breakthrough Moment

You are not defined by your past struggles. Like Angela, you have the power to break free from learned helplessness and rewrite your story. Each step forward, no matter how small, is proof of your strength.

Remember: your ability to succeed isn't determined by your past—it is fueled by the actions you take today. With every small win, you are proving to yourself that transformation is possible. You've got this.

The 10 Characteristics of Lasting Success

Success leaves clues, and people who achieve lasting health share certain traits. Here are ten characteristics you can cultivate on your journey:

1. **Clarity:** They know what they want and why it matters.
2. **Commitment:** They follow through on their goals, even when it's hard.
3. **Resilience:** They bounce back from setbacks stronger than before.
4. **Adaptability:** They're willing to pivot and try new strategies.
5. **Self-Awareness:** They understand their triggers and patterns.
6. **Focus:** They prioritize their health without distractions.
7. **Gratitude:** They appreciate progress, no matter how small.
8. **Patience:** They trust the process and avoid rushing results.
9. **Support-Seeking:** They surround themselves with people who lift them up.
10. **Certainty:** They believe, without a doubt, that success is possible.

These traits aren't innate—they're cultivated through consistent effort. Which of these characteristics resonates with you?

Which one will you focus on building first?

Visualization and Emotional Conditioning: Training Your Brain for Success

Have you ever noticed how athletes visualize their success before it happens? A sprinter imagines crossing the finish line, feeling the adrenaline, and hearing the crowd's cheers. This isn't just motivation—it's mental conditioning. By visualizing success, athletes prime their bodies and minds to make it real.

You can use the same technique. Each day, close your eyes and vividly imagine yourself living in your healthiest, happiest body. Picture yourself walking confidently, feeling energized, and smiling as you notice your progress. Engage all your senses—how does it feel, look, and sound?

Emotional conditioning is key here. When you attach positive emotions to your vision, your brain becomes invested in making it happen. The more vividly you can see and feel your success, the more likely you are to achieve it.

See it. Feel it. Achieve it!

Transformational Vocabulary: How the Words You Use Shape Your Reality

Have you ever noticed how the words you speak—both aloud and in your mind—shape your experience? Your vocabulary isn't just a tool for communication; it's a mirror of your mindset and a compass guiding your actions. The way you describe your journey, your progress, and even your setbacks can either lift you up or drag you down.

Let's imagine two people facing the same situation. Both are navigating the challenges of eating healthier while balancing the demands of their busy lives.

The first person wakes up and says, "I'll try to eat healthy today." The tone is tentative, uncertain, leaving room for failure. They're already framing the day as a battle they might lose.

The second person wakes up and declares, "I am choosing foods that nourish my body today." It's firm, intentional, and rooted in self-respect. That single statement reinforces their sense of control and purpose.

This is the power of Transformational Vocabulary. By shifting the *way* you talk about yourself and your journey, you're not just changing your words—you're rewiring your brain to see success as inevitable.

Let me geek out here for a minute.

Bear with me—I just find the process of rewiring the brain absolutely fascinating. Not to get too technical, but when you honestly believe the positive transformational words you are telling yourself, something remarkable happens on a neurological level. Neuroscientists call this *Hebb's Rule*: "Neurons that fire together wire together." In simple terms, the more you repeat a thought, belief, or behavior, the stronger the neural connections become. Your brain starts reinforcing that pattern until it becomes second nature.

The human mind is far more moldable than most of us were led to believe. Just as negative thoughts and limiting beliefs can keep us stuck, intentional, positive language can literally reshape the way we think, act, and perceive ourselves. This isn't about feel-good affirmations—it's about using the science of neuroplasticity to create lasting change.

If you're a geek like me, you can check out Dr. Donald Hebb, the Canadian neuroscientist who first introduced this principle in the 1940s, showing how repeated thoughts and experiences physically reshape our brain's neural connections.

Reframing the Language of Failure into the Language of Success

Let's look at some common phrases and how a simple tweak can completely transform your perspective:

Failure: "I can't have that dessert."

The word "can't" creates a sense of restriction and deprivation. It implies that something external is forcing you to say no, which can trigger feelings of rebellion or resentment.

Success: "I don't want that dessert because I value how I feel."

Now, the choice is yours. You're not denying yourself—you're empowering yourself by aligning your actions with your values. It's no longer about restriction; it's about respect for your body and your goals.

Failure: "This is too hard."

This phrase creates a mental roadblock, making challenges feel insurmountable.

Success: "I'm learning to overcome challenges every day."

Here, the focus shifts from difficulty to growth. You are reminding yourself that progress isn't about perfection—it's about learning, adapting, and building strength over time.

Failure: "I messed up again."

This statement is laced with self-blame, reinforcing a narrative of failure.

Success: "I made a choice that didn't serve me, but I'm proud of myself for noticing it and resetting."

With this reframe, you take ownership without shame. You are focusing on awareness and growth, not the mistake itself.

The Neuroscience of Thought Patterns: How Repeating a New Narrative Rewires Your Brain (AKA Neuroplasticity)

Imagine this: You're at a family gathering, and your cousin offers you a slice of her famous triple chocolate cake. You could say, "I can't have that," and feel like you're missing out—or you could say, "I'm choosing not to have that because I've been feeling so good eating lighter, and I want to keep that momentum."

The words you choose in that moment can completely change your experience. One approach leaves you feeling deprived, while the other reinforces your strength and commitment.

Rewriting Your Story

Instead of seeing yourself as someone "struggling to lose weight," start describing yourself as someone "building strength, health, and vitality."

Instead of saying, "I've always failed at weight loss," say, "I'm learning what works for me, and I'm making progress every day."

Your language doesn't just describe your reality—it *creates* it. The more you affirm your ability to grow and succeed, the more your actions will align with that belief.

A Real-Life Example

Meet Susan, one of my clients. She used to say, "I'm terrible at sticking to healthy habits." Every time she missed a workout or indulged in a sugary snack, she would tell herself, "Here I go again, proving I can't do this."

We worked together to shift her internal dialogue. Susan began replacing her self-defeating language with phrases like, "I'm learning to prioritize my health, even when life gets busy," and "Every small step I take is a step toward the person I want to be."

The change in her words led to a change in her mindset. She stopped seeing herself as someone destined to fail and started seeing herself as someone capable of growth and transformation.

A Challenge for You

Take a moment to reflect on the words you use when you talk about your journey. Are they words of empowerment, or are they reinforcing the struggles you are trying to overcome?

Start small. Pick one phrase or thought that you find yourself saying often and reframe it using the language of success. Practice it daily until it feels natural.

Because here's the truth: Your words are your blueprint, and you have the power to author a story of strength, growth, and success. Let's make your story a great one.

A Message to GLP-1 Users: You've Always Been in Control

If you have relied on GLP-1s, it's natural to feel apprehensive about maintaining your progress without them. But remember this: The medication didn't do the work—you did. It didn't prepare the healthy meals, go on the walks, or choose to prioritize your well-being. You did.

Belief in your ability to thrive without GLP-1s is the key to freedom. When you trust yourself and commit to the habits you've built, you'll discover that the strength to succeed has always been within you.

Looking Ahead

In the next chapter, we will explore how to solidify your progress by creating sustainable habits that honor your unique biology and lifestyle. Remember, belief is your starting point. When you believe in yourself and your ability to succeed, you create the foundation for a lifetime of health, freedom, and vitality.

Let's continue building that foundation—together.

Your Turn: Integrate & Empower

Mini Exercise: Strengthening Your Belief, Transforming Your Words, and Visualizing Success

> *Want to go deeper? A link to the full Whole-BEING Empowerment Workbook is available in the Resources section of this book.*

Step 1: Strengthen Your Belief in Your Success

Your success starts with belief. When you trust yourself, your actions follow.

📖 **Journal Prompt:**

- Recall a time when you achieved something you once thought was impossible. How did you push through doubt?
- How can you apply that same determination to your health journey?

📌 **Action Step:**

- Write one statement affirming your success. Example: *"I am strong, capable, and fully in control of my health journey."*

✅ **Daily Action:**

- Start your morning with a glass of water as a physical reminder of your commitment to yourself.

Step 2: Transform Your Words, Transform Your Mindset

The words you use shape your experience. Let's shift them in your favor.

📖 **Journal Prompt:**

- What is one limiting belief you often say to yourself? (Example: *"I've always struggled with my weight."*)
- How can you reframe it into an empowering belief? (Example: *"I am learning to nourish my body in a way that supports my goals."*)

📌 **Action Step:**

- Choose one phrase you will stop saying and replace it with a more empowering version.

✅ **Daily Action:**

- Each time you catch yourself using limiting language, pause and replace it with your new empowering statement.

Step 3: Visualize Your Future Success

Your brain responds to what you consistently picture—let's train it for success.

📖 **Journal Prompt:**

- Close your eyes and visualize yourself at your healthiest. How do you feel? What do you see?
- What is one action you can take today to move closer to that vision?

📌 **Action Step:**

- Spend one to two minutes each morning visualizing your success and feeling the emotions tied to it.

✅ **Daily Action:**

- Take a short walk and reflect on your progress—movement reinforces confidence and commitment.

✅ **Final Reflection & Actions to Reinforce Your Growth:**

- What is the biggest insight you gained from this chapter?
- How will you apply it to your journey moving forward?

📌 **Daily Success Habits:**

✓ Continue your daily 15-minute walk—movement reinforces belief in your progress.

✓ Keep your transformational statement visible—repeat it every morning.

Part 4

Sustainable Habits for Whole-BEING Wellness

Mastering Nutrition, Movement, Sleep, and Stress for Lifelong Health

*"Wellness isn't a weekend plan.
It's who you become, one aligned choice at a time."*

— Holli Bradish-Lane

Short-term diets don't work. Sustainable habits do. This section is where you learn how to nourish your body, move in ways that feel good, and build the key pillars of long-term well-being.

We'll cover nutrition, movement, stress resilience, and sleep—the four non-negotiables for lasting health. You will learn how to balance macronutrients to control hunger, use exercise as a tool for metabolism, and optimize your sleep and stress for total-body wellness.

By the end of this section, you will have a lifestyle that fuels you—not a restrictive plan that drains you. You'll feel stronger, more energized, and completely in control of your health.

This is where short-term effort transforms into lifelong success. Let's build the foundation.

Chapter 9

Building Your Nutrition Foundation— The Cornerstone of Lasting Health and Vitality

Introduction: The Power of Food in Your Life

Food is more than just fuel. It's a way to nurture your body, mind, and spirit—a foundation for energy, vitality, and well-being. Yet, in today's world, food has become a source of confusion, guilt, and even frustration.

Food is at the heart of so many of life's most joyful moments. It is the centerpiece of celebrations, from birthdays and weddings to Thanksgiving feasts. It's the way we express love, connection, and care—cooking a warm meal for family, sharing laughter over dinner with friends, or gathering around the table during the holidays. For centuries, food has brought people together, whether at sacred feasts, grand banquets, or simple meals shared at home.

But here lies the paradox: while food has the power to nourish every cell in our bodies and bring us joy, it can also become a way we seek to fill emotional voids, soothe stress, or meet unmet needs for fulfillment. What begins as a celebration can sometimes turn into overindulgence, leaving us feeling disconnected, frustrated, or even ashamed. This tension—between honoring food as nourishment and grappling with its role in overeating—can leave us stuck in a cycle of guilt and confusion.

You have likely tried countless approaches to eating: calorie counting, intermittent fasting, low-carb, or even the latest fad diet. And maybe some plans worked for a while—until they didn't. If you've been on GLP-1 medications, you may have found relief from the constant battle with hunger and cravings. But now, as you consider life beyond medication, it's time to reframe your relationship with food—not as something to fear or fight against, but as a source of nourishment, vitality, and even joy.

Here's the truth: there's no one-size-fits-all solution when it comes to nutrition. What works for one person may not work for another because your body's needs are unique to you.

This chapter is all about helping you build a solid nutritional foundation that aligns with your genetic blueprint, your goals, and your lifestyle. It is not about deprivation or rigid rules. It's about understanding your body, feeding it with intention, and creating habits that make you feel strong, confident, and empowered—all while embracing the joy and connection that food can bring to your life.

Eating for Your Genetic Blueprint

Imagine trying to solve a puzzle with pieces from someone else's set—it just doesn't work. That's what happens when we follow generic diet plans that don't account for our individual biology.

Your DNA holds the key to unlocking a personalized approach to nutrition. For instance:

- Are you more sensitive to carbohydrates or fats?
- How efficiently does your body process caffeine?
- Do you have genetic predispositions that influence cravings or hunger signals?

Understanding your genetic blueprint allows you to make informed choices. You are no longer guessing or following trends—you're creating a plan that works *for you*.

Lisa's Story: True Energy Breakthrough with DNA-Based Nutrition

Lisa had always been active. As a nurse manager at a large university teaching hospital, she was used to the fast pace of hospital life, constantly moving, making critical decisions, and ensuring everything ran smoothly. A year before she reached out to me, she was a floor supervisor, spending entire shifts on her feet, walking miles each day without even thinking about it. But after she was promoted to a managerial role, her daily routine changed—more time at a desk, fewer steps, and a growing sense of exhaustion.

It didn't take long before she started feeling the effects. Her energy plummeted. By the time she got home, she was wiped out, struggling to find motivation for workouts or even simple activities she once enjoyed. On top of that, she had gained 10 pounds despite her efforts to eat "clean." She felt stuck, frustrated, and unsure of what had changed.

When Lisa joined DNAslim, we took a deeper look—starting with her genetic blueprint. What we found surprised her: her body didn't thrive on the high-fat keto diet she had been following. Instead, her DNA revealed that she performed best on a balanced intake of complex carbohydrates and proteins. Ironically, the very thing she had been avoiding—healthy, slow-digesting carbs—turned out to be the key to stabilizing her energy and metabolism.

Armed with this new insight, Lisa adjusted her meals. Within days, she noticed the difference—steadier energy throughout the day, sharper focus at work, and no more mid-afternoon crashes. Within weeks, she felt like herself again, with the stamina to get through her demanding schedule and still have energy left for the things that mattered outside of work.

This kind of insight is powerful. It's not about guesswork. It's about working with your body, not against it. And for Lisa, that changed everything.

Balancing Macronutrients for Hunger Control

Hunger is one of the most common—and frustrating—challenges on the journey to a healthier body. Those persistent cravings, gnawing feelings of emptiness, or that insatiable urge to snack can derail even the most determined efforts. If you've been on GLP-1 medications, you may have experienced the powerful effect of reduced appetite and quieted food noise—making hunger feel like less of a battle. But as you transition away from medication, or if you're looking to optimize your nutrition long term, the key lies in balancing macronutrients—proteins, fats, and carbohydrates—to create meals that nourish your body and keep you satisfied.

When your macronutrients are balanced, you set your body up for steady energy, fewer cravings, and improved focus throughout the day. Each macronutrient serves a unique purpose in hunger control and overall wellness:

The Role of Protein: The Hunger Tamer

Protein is the cornerstone of any meal aimed at managing hunger. It helps regulate appetite by stimulating the release of hormones like **GLP-1** (yes, the same pathway that GLP-1 medications work through) and **PYY (Peptide YY)**, a gut hormone that signals fullness to your brain and helps reduce overall food intake. Beyond hunger control, protein is critical for repairing and building lean muscle, which can boost your metabolism and help you burn more calories even at rest.

> Sources: Eggs, chicken, turkey, fish, Greek yogurt, tofu, legumes, and protein powders.

Healthy Fats: The Satiety Superstars

Fat has been unfairly demonized in the past, but it is one of the most important nutrients for keeping you satisfied. Healthy fats provide long-lasting energy, slow digestion, and make meals more satisfying. They also help with the absorption of fat-soluble vitamins like A, D, E, and K, supporting your overall health.

> Sources: Avocado, nuts, seeds,
> olive oil, fatty fish (like salmon), and coconut oil.

Complex Carbohydrates: The Steady Energy Source

Complex carbohydrates are a valuable source of energy that fuel both your brain and your body. Unlike simple carbs (like white bread or sugary snacks) that cause rapid blood sugar spikes and crashes, complex carbs release energy slowly, helping to prevent cravings and maintain focus. They are also a reliable source of fiber, which aids digestion and enhances satiety.

> Sources: Quinoa, sweet potatoes,
> whole grains, brown rice, oats, fruits, and vegetables.

Common Pitfalls When Balancing Macronutrients—And How to Fix Them

When your meals are out of balance, hunger can creep in quickly:

- A meal high in carbs without enough protein or fat may cause your blood sugar to spike, leaving you energized briefly but starving and irritable later.
- A meal that's only protein may leave you unsatisfied, as fat and carbs play important roles in satiety and energy.

- Skipping meals entirely can lead to overcompensating later, triggering cravings for high-calorie, low-nutrient foods.

Balancing your macronutrients allows you to create meals that keep your energy steady, your cravings under control, and your metabolism humming.

A Practical Example: Rebalancing a Typical Meal

Let's put this into perspective:

Imagine a typical "healthy" lunch—say, a salad with greens and a low-calorie dressing. While it might sound virtuous, this meal is missing key components. By 3 p.m., your energy is crashing, and you're eyeing the leftover donuts in the breakroom.

Now let's rebalance that meal:

- Add **grilled chicken** or **tofu** for protein to keep you full.
- Top the salad with **sliced avocado** or a sprinkle of nuts for healthy fats to create satiety.
- Toss in a serving of **quinoa**, **chickpeas**, or **sweet potato chunks** for complex carbs that provide long-lasting energy.

With this balanced meal, you will feel satisfied for hours and avoid the temptation of sugary snacks or empty calories.

Tips for Mastering Macronutrient Balance

1. **Use the "Balanced Plate" Method:** Divide your plate visually:
   - ⅓ Protein
   - ⅓ Vegetables or Fiber-Rich Carbs (e.g., leafy greens, sweet potatoes, or quinoa)
   - ⅓ Healthy Fats (e.g., avocado slices, olive oil, or a handful of nuts)

2. **Plan Ahead:** If you wait until you are hungry to decide what to eat, you're more likely to grab the easiest or most indulgent option. Prep balanced meals or snacks in advance to set yourself up for success.
3. **Watch Portions, Not Just Macronutrients:** While balance is crucial, portion size matters too. Even healthy foods can add up in calories if you're not mindful. Aim for palm-sized servings of protein, a thumb-sized serving of fats, and a cupped-hand portion of carbs.
4. **Adjust Based on Your DNA and Lifestyle:** Your genetics play a role in how your body processes macronutrients. For example, if your DNA reveals a sensitivity to carbohydrates, you might find better success with slightly higher protein and fat ratios. Alternatively, if you are more genetically predisposed to burn carbs efficiently, you can include them more freely in your meals.

A Note for GLP-1 Medication Users: How to Maintain Hunger and Craving Control

If you've been on GLP-1 medications, you may have noticed how they helped quiet the hunger and cravings that once felt overpowering. Now, as you taper off these medications, macronutrient balance becomes your new secret weapon. A balanced plate won't just fuel your body—it will also help you sustain the hunger and craving control you experienced on GLP-1s.

The beauty of balancing macronutrients isn't just about the science—it's about the satisfaction. A meal that is thoughtfully balanced not only nourishes your body but also leaves you feeling energized, empowered, and confident that you're taking steps to support your health and well-being.

By learning to balance macronutrients, you're not just managing hunger—you're mastering it. And with that mastery comes freedom: freedom from cravings, from deprivation, and from the rollercoaster of energy highs and lows. Every balanced meal you create is a step toward sustaining your success and building a lifestyle you love.

Managing Triggers with Food Awareness

Let's be honest: most of us don't overeat because we are physically hungry. We eat because something—a trigger—nudges us toward food, whether it's the smell of a favorite snack, a tough day, or even sheer boredom. These triggers fall into three main categories:

- **External Cues:** The smell of fries as you drive past a fast-food restaurant, the sight of cupcakes at a birthday party, or the sound of a soda being poured on a commercial.
- **Mental Cues:** Thoughts like, "I've earned this after today," or "I've had a tough week—why not treat myself?"
- **Physical Cues:** Genuine hunger, like a growling stomach or that drained, low energy feeling when your body needs fuel.

The Science Behind Food Cravings

In today's world, external cues are everywhere, and they are designed to grab your attention when you are most vulnerable. Picture this:

It is 6 p.m. You're crawling through traffic after a long, stressful day at work. Your stomach feels like it's tied in knots—not entirely from hunger, but from the tension that has been building all day. You are tired, mentally drained, and dreaming of getting home to unwind.

But then, it hits you.

The golden glow of a fast-food restaurant's sign pierces the evening sky, promising a moment of comfort. And it doesn't stop there. As you continue driving, the next block offers another temptation: a pizza delivery car zooms by, practically inviting you to imagine biting into a hot, cheesy slice. Minutes later, the unmistakable aroma of fried chicken wafts through your car vents as you pass yet another restaurant.

By the time you're halfway home, you've passed four fast-food chains. Each one whispers, "Why not? It's easy, it's quick, and you deserve it." And in that moment—hungry, tired, and stressed—it's easy to justify pulling into the drive-thru.

How Marketing and Social Cues Trick You into Eating More

This isn't just a lack of willpower. Food marketers are experts at capitalizing on your vulnerabilities. They know how to create scents, lighting, colors, and slogans that pull you in. Add stress, fatigue, or hunger into the mix, and suddenly the idea of cooking a balanced meal feels like climbing a mountain.

Here's the thing: these triggers are not your fault, but they are your responsibility to manage. And the first step in doing that is recognizing what is happening.

The Genetics of Food Triggers: Why Some People Are More Susceptible

Have you ever wondered why certain people can walk past a bakery and not give it a second thought, while others feel an almost magnetic pull toward the smell of fresh bread? The answer might be in your DNA.

Certain genetic variations make some individuals more sensitive to food-related cues—things like smells, sights, sounds, and even memories tied to eating. Research shows that genes involved in **dopamine regulation** (the brain's reward system) and **ghrelin production** (the hunger hormone) play a significant role in food cravings and impulse control.

For example, certain people have genetic variants that make them more reactive to food-related stimuli. Their brains light up more intensely in response to the sight or smell of food, making cravings harder to ignore.

Others may have variations in the **FTO gene**, which is linked to increased appetite and a stronger emotional response to food cues.

This doesn't mean you're doomed to struggle with cravings forever—awareness is power. If you know that your genetics make you more susceptible to food marketing, emotional eating, or environmental cues, you can take proactive steps to **interrupt the cycle** and make mindful choices.

Practical Strategy: If you know you are overly sensitive to food triggers, you can create personal safeguards, such as avoiding grocery shopping when you're hungry, using the Tomorrow Technique for cravings, or physically removing yourself from environments that make resisting temptation difficult.

Building Awareness and Resilience Around Food Choices

Triggers only have power when they are operating in the background, influencing you without your awareness. But once you recognize them for what they are, you can disrupt the cycle.

Start by asking yourself a few key questions in those moments:

- **Am I truly hungry?** If the answer is no, then it's a mental or external cue driving your craving.
- **What triggered this urge?** Did you see a fast-food billboard? Smell something enticing? Did a stressful moment at work leave you searching for comfort?
- **What do I really need right now?** If you're tired, you need rest. If you're stressed, perhaps a deep breath or a short walk could help.

Let's revisit that drive home. Imagine the same scenario, but this time, you're prepared.

As you approach the first fast-food sign, you acknowledge the craving. "Of course, I'm tempted," you think. "That is their job—to make me want it. But I know this is not what my body really needs."

Instead of giving in, you reach for the healthy snack you packed in your bag this morning—a handful of nuts or a protein bar. You take a deep breath, sip water, and focus on how proud you'll feel when you walk into your home later and enjoy the balanced dinner you've planned.

You pass the next few restaurants with a new sense of clarity. Yes, the signs are still glowing, and the smells are still wafting. But they no longer have control over you.

Tools for Long-Term Success: Meal Prep, Distractions, and Boundaries

Here are tools you can use to take back control:

1. **Prepare in Advance:** Keep healthy snacks like fruit, nuts, or a protein bar with you, especially during times when you know you'll be vulnerable—like your drive home or after a long meeting.
2. **Change Your Route:** If you know certain streets or areas are packed with temptations, find an alternate way home. Out of sight, out of mind.
3. **Pause Before Acting:** When you feel the pull of a trigger, pause. Take three deep breaths and ask yourself, "Will this choice bring me closer to or further from my goals?"
4. **Visualize Success:** Picture yourself saying no to the drive-thru and walking into your home to prepare a delicious, balanced meal. Imagine the pride and energy you'll feel after making that choice.

Food is meant to nourish and bring joy, but in today's world, it is often used as a quick fix for emotional or mental discomfort. The problem isn't the food itself—it's the way we've been conditioned to respond to it. Fast food isn't inherently evil, but when it becomes a default reaction to stress or fatigue, it works against us instead of for us.

By building awareness and preparing yourself to navigate triggers, you're not just resisting cravings—you're reclaiming your power. Each time you recognize a trigger and choose differently, you're rewriting your story, one decision at a time.

Remember, you are in control—not the brightly lit signs, not the seductive smells, and not the marketing campaigns designed to exploit your vulnerabilities. With awareness and intention, you can break free from the grip of external cues and create a healthier, more empowered relationship with food.

Breaking Patterns of Emotional Eating

Emotional eating is a common challenge for many people, and it often feels like an endless cycle. If this sounds familiar, you're not alone. While we have touched on some of these ideas in earlier chapters, it's worth repeating—because understanding and breaking this pattern is foundational to building a healthier relationship with food.

Let's break it down:

1. **Trigger:** You feel an emotion—stress, boredom, loneliness, or even celebration.
2. **Urge:** You crave something comforting, like chips, chocolate, or ice cream. The thought of eating feels like a quick fix for how you are feeling.
3. **Behavior:** You eat the food, often without even realizing just how much you're consuming.
4. **Consequence:** Moments later (or after the binge), guilt, regret, or frustration sets in.

This cycle can feel impossible to escape, but here's the truth: **it is entirely possible to break free.**

Practical Techniques to Break the Chain of Emotional Eating

In previous chapters, we explored tools like The Scramble Technique and Pattern Interrupts, and they're powerful for disrupting the cycle of emotional eating. Let's build on those strategies with additional actionable steps:

1. Pause Before You Act

When the urge to eat strikes, pause and take three deep breaths. In that moment, ask yourself:

- "Am I physically hungry, or am I looking for comfort?"
- "What do I truly need right now?"

This pause helps you disrupt the autopilot response and shift into mindful decision-making. Often, you'll realize that hunger isn't driving your craving—it's an emotional need.

For example, let's say you have had a long, exhausting day. You open the cabinet and spot a bag of red licorice. Before reaching for it, pause. Instead of eating, you might discover that what you really need is a moment to decompress. Perhaps a cup of herbal tea, a warm bath, or a quick walk could soothe you better than sugar ever could.

2. The Tomorrow Technique: A Simple but Powerful Pause

Remember, this technique is deceptively simple but highly effective. When a craving hits, tell yourself, **"If I still want it tomorrow, I can have it."**

The beauty of this approach lies in its flexibility. You're not denying yourself—you're just delaying the decision. More often than not, the craving passes.

Imagine opening the refrigerator and seeing the leftover raspberry pie from last night's dinner. You feel the urge to grab a slice, but instead, you say to yourself, **"That pie will still be here tomorrow. I don't have to eat it now."** By postponing the decision, you give yourself time to evaluate whether the craving is worth acting on.

3. Plan for Vulnerable Moments

We all have times of day or situations when we are most vulnerable to emotional eating. For many people, it's the evening—after a stressful day, when the house is quiet and the cravings creep in.

Here is how to set yourself up for success:

- **Prepare Healthy Alternatives:** Have satisfying, nutrient-dense snacks ready, like Greek yogurt with berries, a handful of almonds, or cut-up veggies with hummus.
- **Create a Go-To Distraction List:** Write down activities that can shift your focus away from food, such as reading a book, journaling, taking a walk, or calling a friend.
- **Close the Kitchen:** Set a mental boundary around food after a certain time. For example, you might decide that the kitchen "closes" at 7 p.m., signaling to your brain that the eating window for the day is over.

4. Reframe Emotional Eating

Rather than beating yourself up for emotional eating, approach it with curiosity and self-compassion. Emotional eating isn't a failure—it's a coping mechanism you've developed over time. By understanding it, you can learn to replace it with healthier ways of coping.

For instance, if you're feeling stressed, instead of diving into a bag of chips, try journaling about what is bothering you. Or, if you're bored, engage in an activity that stimulates your mind, like solving a puzzle or learning a new skill.

Breaking the Paradox of Food and Emotion

Food is deeply emotional, and that's okay. It's woven into celebrations, family traditions, and memories. The goal isn't to strip food of its emotional significance—it's to create a healthier relationship with it. Food can still bring joy and comfort, but it shouldn't be your only source of fulfillment.

As Geneen Roth explores in *When Food is Love*, emotional eating isn't about hunger—it's often an attempt to fill a deeper need. When food becomes a stand-in for love, comfort, or self-acceptance, it can create a cycle where eating is less about nourishment and more about soothing wounds from the past. Recognizing this is the first step to breaking free. True nourishment comes from more than what's on your plate—it comes from how you care for yourself, set boundaries, and cultivate joy beyond food.

Remember: every time you choose a different response to a trigger, you're rewriting your story. Emotional eating doesn't have to control you—you have the tools to take back control and make choices that align with the life you're building.

Ready to Understand Your Eating Triggers? Take the Emotional Eating Quiz

Emotional eating isn't just about hunger—it's often about patterns, triggers, and emotions you might not even realize are at play. This short quiz will help you uncover what is really driving your cravings and give you personalized insights to break free.

- **Take the Emotional Eating Quiz now** → *www.ironcruciblehealth.com/craving-clarity-quiz*

By taking the quiz, you'll also receive a **FREE Emotional Eating Guide** with practical steps to regain control—because food should be a source of nourishment, not stress.

The journey may not be perfect, but each step forward is progress. Whether you are practicing the Tomorrow Technique, pausing before acting, or planning for vulnerable moments, you're breaking the cycle and creating new patterns that serve you.

Take it one moment at a time—you've got this.

The Influence of Society and Advertising on Your Eating Habits

We live in a world designed to make you overeat. From endless commercials for fast food to grocery store layouts that strategically place sugary snacks at eye level, everything is engineered to tempt you. Marketers know how to tap into your emotions and subconscious triggers, making you feel hungry even when you're not.

Remember those Girl Scout cookies we talked about a few chapters back? Or that Kohl's "VIP 40% off" email that made you feel like you were part of an exclusive club? Those are classic examples of how scarcity and exclusivity create urgency—you feel like you must act fast because the opportunity won't last.

The same psychological tactics are used in food advertising to get you to overconsume:

- **Limited time offers** create urgency, convincing you to "act now."
- **Bright, colorful packaging** grabs your attention, especially when you are hungry.
- **Slogans like "You deserve a break today"** tap into your emotions, making indulgence feel like self-care.

Here is a real-world example that hit me like a lightbulb moment...

On a recent weekend, I settled in to watch the National Football League (NFL) playoffs. While my beloved Denver Broncos didn't make it all the way this year (hey, there's always next season!), I was still pumped for the Super Bowl showdown. But midway through the playoff game, I noticed something strange—I started feeling hungry, even though I had eaten a balanced meal just a few hours earlier.

Then it clicked: it wasn't hunger. It was the commercials.

In just one hour of watching, I counted **44 commercials. Fifteen of them—more than a third—were for food and drinks.** Pizza, fried chicken, burgers, beer, soda—their messages were everywhere. The ads were not just targeting my attention; they were targeting my appetite, trying to convince me that I needed to eat. And here's the kicker: even weight-loss drug commercials like GLP-1s were sandwiched between food ads, adding another layer of irony.

These ads are designed to hijack your decisions. They use vivid imagery, comforting language, and emotional appeals to trick your brain into thinking, "I need that."

Building Resilience Through Awareness

The first step to fighting back against these subtle influences is awareness. **Recognizing that these messages are engineered to manipulate your cravings is empowering.** The more mindful you are of how advertising impacts your thoughts, the less likely you are to act on false hunger cues.

- **Pause and check in with yourself.** Next time you feel an urge to snack—especially after seeing a food ad—ask yourself, "Am I actually hungry, or is this just a reaction to what I saw?"

- **Create boundaries around triggers.** For instance, if food commercials tempt you, use those ad breaks as a cue to do something else—stand up, grab a glass of water, or take a moment to stretch.
- **Remind yourself who is in charge.** You control your body, not the marketing machine. Tell yourself: **"I choose what nourishes me."**

Remember, the world may be designed to challenge your resolve, but you are stronger than the marketing machine. Whether it's resisting the lure of fast food on your drive home or skipping the sugary snacks in the grocery aisle, every mindful decision you make is a win.

Awareness is power. Strength is yours.

Breaking old patterns and creating new ones takes time, patience, and self-compassion. Again, it's not about being perfect—it's about being consistent and learning from your experiences.

Every time you choose a balanced meal over fast food or pause before giving in to a craving, you're strengthening your ability to make intentional, aligned choices. You're building resilience—not just in your eating habits, but in your entire approach to health and well-being.

Closing Encouragement

"Health isn't built in bursts—it's built in rhythms."

— Holli Bradish-Lane

Your nutrition foundation is about so much more than what you eat. It's about understanding yourself—your triggers, your patterns, and your needs—and creating a plan that supports your goals.

Remember: every choice is a step towards the vibrant, healthy life you are creating. Celebrate your progress, stay curious, and keep building on the foundation you have started. Your body and mind are on this journey together, and with each aligned choice, you are showing yourself just how capable you are.

Next up, we will explore how movement and fitness fit into your unique blueprint for whole-being wellness. Let's keep building your momentum!

Your Turn: Integrate & Empower

Mini Exercise: Building Your Nutrition Foundation

Want to go deeper? A link to the full Whole-BEING Empowerment Workbook is available in the Resources section of this book.

Step 1: Identify Your Triggers and Patterns

Understanding your eating patterns is the first step to creating lasting change.

Journal Prompt:

- Think back to the last time you ate something out of impulse or emotion rather than true hunger.
    - What was happening at that moment? (e.g., stress, boredom, a food ad, a smell)
    - Were there any external triggers? (e.g., a fast-food sign, social pressure, seeing a snack in your pantry)
    - How did you feel before, during, and after eating?

📌 **Action Step:**

- For the next three days, write down any cravings you experience.
- Note whether they were triggered by true hunger or an external/mindless cue.

Step 2: Balance Your Meals for Lasting Satisfaction

Balanced meals help keep cravings at bay and prevent overeating.

📖 **Journal Prompt:**

- Reflect on your typical meals. Do they include a balance of **protein, healthy fats, and complex carbs**?
- Do you feel satisfied after eating, or do you experience cravings and energy crashes later?

📌 **Action Step:**

✓ For your next meal, use the *Balanced Plate Method:*

- ⅓ **Protein** (e.g., chicken, tofu, Greek yogurt)
- ⅓ **Vegetables or Fiber-Rich Carbs** (e.g., leafy greens, quinoa, sweet potatoes)
- ⅓ **Healthy Fats** (e.g., avocado, olive oil, nuts)

✓ Track how you feel after eating. Are you satisfied longer?

Step 3: Disrupt the Cycle of Emotional Eating

Emotional eating is often a learned habit, but it *can* be rewired.

📖 **Journal Prompt:**

- What emotions most often lead you to eat when you are not hungry? (e.g., stress, loneliness, celebration)
- What non-food strategies could help you manage these emotions instead?

📌 **Action Step:**

✓ Next time you feel the urge to eat for emotional reasons, **pause for three minutes** and try one of these alternatives:

- Drink a glass of water or herbal tea.
- Take five deep breaths.
- Go for a five-minute walk.
- Write down how you are feeling before making a food decision.

☑ **Final Reflection:**

- What is one insight from this chapter that changes how you think about food?
- What is one small shift you can make today that will support your long-term success?

Chapter 10

Fitness for Your DNA and Lifestyle

Moving Your Way to Sustainable Weight Loss

"Eat to nourish. Move to express. Rest to restore. That's alignment."

— Holli Bradish-Lane

Introduction: Movement That Matches Your Life

Exercise isn't about punishment or perfection—it's about empowerment. The goal is to move in ways that nourish your body, honor your DNA, and fit seamlessly into your lifestyle. Whether you are a gym enthusiast, a brisk walker, or someone who hasn't laced up sneakers in years, this chapter is all about helping you find joy in movement and making fitness sustainable.

For many, fitness feels overwhelming—another item on an endless to-do list. If you have been on GLP-1 medications, you may have experienced the benefit of reduced hunger or the energy to stay active. But as you consider life beyond the medication, movement will become an even more essential part of your journey. Not as a task to check off, but as a gift that reconnects you with your body and helps you maintain your progress naturally.

Movement is about far more than burning calories. You're building strength, boosting your mood, and supporting long-term health. The best part? By understanding your unique DNA blueprint, you can tailor fitness to work with your body—not against it—making it feel less like a battle and more like a partnership.

Overcoming Early Negative Experiences

For some, the word "exercise" carries a weight of its own—one filled with negative associations that stem from early experiences. Perhaps you have felt like you weren't "athletic enough" or dreaded PE class because of awkward or even humiliating moments.

I remember being that shy, skinny kid in elementary school, cringing at the sight of dodgeballs lined up in the gymnasium. I wasn't fast or strong, and the thought of a bigger boy launching a ball at me—landing with a stinging smack on my thigh or stomach—was terrifying. Dodgeball did not feel like a game; it felt like survival.

Those early experiences shaped my view of exercise. I saw it as something to endure, not enjoy. It wasn't until much later in life that I realized movement didn't have to be competitive, painful, or something I dreaded. It could be something personal, empowering, and even joyful.

If your relationship with movement was shaped by similar experiences, know this: it's not too late to rewrite the narrative. Exercise isn't about proving yourself to anyone else—it's about discovering what feels good and works for you.

Why Movement Is More than Just Burning Calories

Exercise is one of the most powerful tools in your arsenal—not just for weight loss, but for creating a vibrant, healthy life. But let's make one thing clear: movement is not punishment for what you ate. It is not about earning your dinner or burning off dessert.

It's about *enhancing* your life in ways that go far beyond the physical. Here is what regular movement can do for you:

- **Reinforce Your Progress Beyond GLP-1s:** Movement helps you maintain your weight naturally, supporting your progress even after GLP-1 medications.
- **Boost Energy:** Ever notice how a short walk can make you feel refreshed? That's because movement improves circulation and oxygen delivery to your cells, leaving you more energized.
- **Elevate Your Mood:** Physical activity releases endorphins—those feel-good chemicals that reduce stress and anxiety while lifting your spirits.
- **Build Confidence:** There's nothing like the sense of accomplishment that comes from finishing a workout, mastering a yoga pose, or simply showing up for yourself.
- **Support Longevity:** Regular movement reduces your risk of chronic illnesses like heart disease, diabetes, and even depression.

Movement is a form of self-care, a way to honor the incredible body with which you've been entrusted. This body—your one and only—carries you through every experience, every challenge, every joy. When you begin to see movement as an act of respect and gratitude for all it does, it stops feeling like a chore and becomes an essential part of your well-being.

Reframing Exercise: The Power of Movement Over Rigid Workouts

Many people don't think they're "fit enough" to start exercising, or they feel they need to look a certain way before walking into a gym. That's a mindset I want to help you overcome.

Fitness is not about keeping up with anyone else or achieving perfection. It is about progress—meeting yourself where you are today and taking small, consistent steps forward.

You are not just moving for your health; you're moving to reclaim your confidence, celebrate your strength, and create a sustainable lifestyle that supports your goals. Whether you're rolling out a yoga mat in your living room, taking a brisk walk in your neighborhood, or dancing in the kitchen while making dinner, every bit of movement counts.

This chapter will help you break free from old, limiting beliefs about exercise and discover a way of moving that feels as unique and free as you are.

Personalized Exercise Based on Your DNA Blueprint

Not all fitness plans work for everyone. Your genetic makeup plays a significant role in how your body responds to exercise and understanding this can help you design a fitness routine that's not only effective but also enjoyable and sustainable.

Think of it this way: If you are built for endurance but keep forcing yourself through grueling sprints, or if your body thrives on strength training but you're stuck in a spin class, you might feel frustrated or burned out. When you align your exercise habits with your genetic blueprint, fitness becomes less of a struggle and more of a flow.

Genetic Factors That Influence Exercise Response

Optimal Exercise Type – Your DNA can reveal whether your body is better suited for power (weightlifting), endurance (running, cycling), or a mix of both.

Recovery Needs – If your genes indicate slower recovery, you will benefit from rest days and active recovery like yoga or stretching.

Injury Risk – Certain genetic markers indicate a predisposition to joint issues or connective tissue weaknesses, making low-impact exercises like swimming a smarter choice.

Caffeine Response – If your DNA suggests caffeine sensitivity, that pre-workout espresso might do more harm than good.

Dana's Story: Energizing Fitness Revelation

Dana, one of my clients, always started fitness programs with enthusiasm but never stuck with them. Boot camp-style workouts felt exhausting, and she would burn out within weeks. She thought she just lacked discipline.

Through DNA testing, we uncovered the real issue: Dana's body wasn't wired for high-intensity interval training (HIIT). Instead, her genetic profile showed she thrived on moderate-intensity strength training and recovery-focused activities like yoga and Pilates.

This simple shift changed everything. Dana stopped feeling defeated by workouts she didn't enjoy and started embracing a routine that worked with her body instead of against it. With this personalized approach, she stayed consistent, felt energized, and finally built a fitness habit that lasted.

Move Your DNA: The Power of Everyday Movement

When we think of fitness, we often picture structured workouts—sweating through a spin class or hitting the gym. But as Katy Bowman emphasizes in her book *Move Your DNA*, movement isn't just about exercise. It is about how we use our bodies throughout the day, every day.

Our modern lives are alarmingly sedentary. We spend hours sitting at desks, driving, or scrolling through our phones, which leaves our bodies craving the natural, varied movement they evolved to perform. The result?

Chronic pain, stiffness, and health issues rooted in lack of mobility—not just lack of exercise.

Bowman's philosophy shifts the focus from isolated workouts to a more holistic approach: weaving movement into your daily routine. Walking to the store instead of driving, gardening, taking the stairs, stretching while watching TV—all these small, natural movements add up.

This perspective complements the idea of personalized fitness. Instead of focusing solely on the gym or specific workouts, think about ***movement as a way of life***. It's not about forcing yourself into rigid routines but about reconnecting with your body and honoring how it was designed to move.

Combining DNA Insights with Everyday Activity

By combining what you learn from your DNA with Bowman's concept of daily movement, you can create a lifestyle that prioritizes health in an intuitive and sustainable way. Imagine incorporating exercises tailored to your genetic strengths while also making small, meaningful shifts to keep your body active throughout the day.

For example, if your DNA suggests a higher need for endurance activities, you might:

- Go for a brisk walk each morning.
- Take short walking breaks during the day to avoid long stretches of sitting.
- End your day with a calming yoga session.

If your genetic profile shows a preference for power-based movements, you might:

- Integrate bodyweight strength exercises while waiting for your coffee to brew.

- Lift light weights during short, focused sessions three to four times a week.
- Incorporate functional movements into daily chores, like squatting to pick up laundry or stretching while unloading groceries.

The beauty of combining genetic insights with Bowman's philosophy is that it removes the "all-or-nothing" mentality around fitness. You are no longer confined to rigid workout programs; instead, you're building a life full of movement and intention.

Your body is unique, and so is your journey. By aligning your fitness habits with your DNA and embracing movement as part of your daily life, you will create a foundation for strength, vitality, and long-term health.

Making Movement Fit Your Life

We live busy lives. Between work, family, and countless responsibilities, finding time to move can feel impossible. But fitness doesn't have to mean spending hours at the gym. The best exercise is the one you will actually do.

Tips for achieving movement:

1. **Start Small:** You don't need a perfect workout routine to start. Even 10 minutes a day adds up over time. Just move.
2. **Be Flexible:** Can't make it to the gym? Do bodyweight exercises at home, take a brisk walk during lunch, or stretch before bed. Movement isn't about location—it's about intention.
3. **Combine Movement with Joy:** Choose activities you genuinely enjoy. Dance, hike, swim, garden—movement comes in many forms.
4. **Make It a Ritual:** Anchor your exercise to something you already do, like walking after dinner or doing yoga before your morning

coffee. Consistency is easier when fitness becomes a part of your daily rhythm. So, pencil it in to your schedule!
5. **Use Movement as a Stress-Relief Tool:** Instead of reaching for food when stressed, take a short walk, do a few stretches, or even try deep-breathing exercises to recenter yourself.

Staying Consistent: The Power of Motivation and Accountability

Starting is one thing—sticking with it is another. We all have moments when motivation fades, when life gets busy, or when old habits try to creep back in. That's why lasting success isn't about willpower—it's about having the right strategies in place to keep yourself accountable and inspired.

1. Redefine Motivation: It is Not About "Feeling Like It."

A common misconception is that successful people are always motivated. The truth? They don't rely on motivation alone—they rely on habits, discipline, and accountability.

Motivation comes in waves. Some days, you'll feel unstoppable. Other days, you will want to skip your workout or reach for comfort food. That's normal. The key is having a system in place for those moments when motivation is low.

Instead of asking, "Do I feel like working out?" ask, "Will I be proud of myself if I do?" Commitment beats motivation every time.

2. Set "Non-Negotiables" for Your Fitness Routine

Instead of relying on willpower, build non-negotiable habits—small, realistic commitments that fit into your life no matter what.

Example: "I will move for at least 10 minutes a day—no matter how busy I am."

Example: "I will do a short walk after dinner instead of scrolling on my phone."

Example: "I will pack workout clothes in my bag the night before."

These small actions create momentum—and momentum keeps you moving forward.

3. Find Your Accountability System

Left to our own devices, it is easy to make excuses. But when someone else is counting on us, we show up differently. That's why accountability is a game-changer.

Ways to Build Accountability:

- Workout with a friend – A walking friend, gym partner, or even an online check-in can help.
- Join a community – A fitness class, group challenge, or even a social media accountability group can provide support.
- Hire a coach – A coach or personal trainer adds structure and holds you to your goals.
- Track your progress – Use a fitness app, journal, or simple calendar to check off completed workouts. Seeing progress visually can keep you engaged.
- Make a public commitment – Tell a friend, partner, or social media circle about your goal. When you say it aloud, you're more likely to follow through.

4. Create a "Why Empower" List

On tough days, your "why" will keep you going. Take a moment to write down three reasons you want to stay consistent.

Example "Why Empower" List:

1. I want to be strong and mobile for my kids and grandkids.
2. I want to feel energized, confident, and in control of my body.
3. I want to prove to myself that I can commit and follow through.

Whenever you want to skip a workout, re-read your list. Let it remind you of the bigger picture.

5. Reward Yourself the Right Way

Instead of focusing on what you cannot do or shouldn't eat, celebrate what you accomplish.

- Non-food rewards: Buy yourself new workout gear, schedule a massage, or set up a fun day trip after hitting a milestone.
- Mini goals and checkpoints: Instead of just aiming for a big end goal, like losing fifty pounds, set mini milestones like completing 10 workouts in a month or increasing your weights. Each small win builds confidence.

Action Beats Excuses

The biggest secret to consistency? Taking action, even when you don't feel like it. Motivation follows action—not the other way around.

Make a deal with yourself: Just start. Lace up your shoes. Do the warm-up. Take the first step. Momentum builds quickly once you begin.

And when in doubt? Come back to your "why." Because this journey isn't just about looking a certain way—it's about building a body, mind, and life you feel proud of.

Maria's Story: Vital Strength Rediscovered

When Maria first came to me, she was cautiously optimistic but uncertain. After years of battling her weight, GLP-1 medications had finally given her a sense of relief. For the first time in a long time, the constant hunger and cravings had quieted, allowing her to shed the pounds that had stubbornly clung for years.

But despite her progress, something felt off. Carrying groceries still left her winded. Climbing stairs was a struggle. She had expected to feel **stronger, healthier, more energized**—but instead, she felt... depleted.

"It's strange," she admitted. "I'm lighter, but I don't feel better. Shouldn't this be easier by now?"

Her frustration was valid. Weight loss alone doesn't automatically lead to strength and vitality—especially when muscle mass is lost along with fat. That is when we turned to her DNA for answers.

Maria's genetic profile revealed several key factors influencing her fitness potential:

- **Higher protein needs** – Her body required more protein to maintain muscle.
- **Best suited for strength training** – She thrived with a mix of resistance workouts and moderate-intensity movement, rather than high-intensity cardio.
- **Slower recovery genes** – Her body needed rest days and activities like yoga or walking to prevent burnout.

With these insights, Maria finally had clarity—and permission to stop forcing herself into exhausting, unsustainable workouts.

Building Strength, Inside and Out

Maria started small, incorporating bodyweight exercises, and gradually increasing her strength. She adjusted her meals to prioritize protein, adding Greek yogurt for breakfast and lean proteins like salmon to her dinners. But the biggest shift wasn't just in her routine—it was in her mindset.

For years, fitness had felt like a punishment. Now, she saw it as **a way to reclaim her strength and energy.** Each workout, each balanced meal, each small victory reinforced her belief in herself.

The results? Maria didn't just maintain her weight loss—she became stronger, more confident, and **finally felt at home in her own body.** She realized that true transformation wasn't just about losing weight; it's about building a body that works for you, not against you.

Her story is proof that **lasting health isn't about deprivation or struggle—it's about empowerment, knowledge, and aligning your choices with what your body truly needs.**

As you reflect on Maria's story, think about how you can create your own fitness and nutrition plan—one that aligns with your body, your goals, and your unique DNA. Your journey isn't about perfection; it is about progress, one small step at a time.

With the right approach, you can reclaim your strength, rebuild your confidence, and thrive in a body that's built for living your best life.

Your Fitness Journey, Your Rules

Again, this isn't about *perfection*—it's about progress. (Have I said this enough so that it is starting to stick?) *Movement* is a celebration of what your body can do, not a punishment for what you ate or how you look.

You don't have to love every workout, but you can find joy in the way movement makes you feel...stronger, more energized, and more alive.

Looking Ahead

In the next chapter, we'll dive into the connection between stress, sleep, and weight loss—because building a healthy lifestyle isn't about food and movement. It's about creating balance and resilience in every part of your life.

Your body is designed to move. Whether it's a walk around the block, a yoga class, a Saturday morning hike, or dancing in your living room, every step forward is a step toward freedom, strength, and vitality. Let's keep moving forward—together.

Movement: A Gift to Your Future Self

Your body is meant to move. Movement isn't a punishment for the extra slice of birthday cake you ate yesterday, nor is it a chore to cross off your endless to-do list. It's a gift—one that builds not just strength but resilience, clarity, and confidence. It's the key to unlocking a life where you feel fully present and vital.

Maria's journey was a reminder of this truth. GLP-1 medications helped her quiet the noise of hunger and cravings, but what truly transformed her life wasn't the number on the scale—it was discovering her strength. She learned that movement was not about keeping up with anyone else; it was about creating a body that supported her, a mind that believed in her, and a heart that felt proud of how far she'd come.

But maybe you're reading this and thinking, *that sounds great... but exercise just isn't for me.* Perhaps your mind flashes back to uncomfortable gym classes or failed attempts to stick to fitness routines in the past. Maybe the very word "exercise" makes you cringe.

Let's redefine what movement means—on your terms.

Redefining Fitness in Your Terms

Forget the image of crowded gyms or punishing workouts. Movement doesn't have to look like anyone else's routine. It is about finding what lights you up, what makes you feel strong, and what helps you reconnect with your body.

For Maria, it started with short walks and evolved into resistance training. For you, it could look completely different. Maybe it's getting on horseback for the first time in years, exploring a new hiking trail, or trying out a beginner yoga video. It could even be parking farther from the grocery store entrance or stretching during your morning coffee.

Fitness doesn't have to be perfect to be powerful. Small steps matter. **Every time you choose movement, you send your future self a message:** *I'm building a stronger, healthier me.*

Finding Joy in Movement That Works for You

Understanding your DNA can take movement to the next level, but you don't need a genetic test to start building momentum. What is most important is listening to your body and finding joy in movement. It's about feeling alive, not just ticking a box.

Even if you've been on GLP-1 medications, progressing to a lifestyle of strength and movement can create freedom. As your body grows stronger, you will realize that the tools you've been using—whether medications, diets, or quick fixes—are not what define you. Your ability to thrive lies in your own actions, your choices, and your determination to keep moving forward.

Practical Steps to Build Momentum

Start where you are, with what you have. If you're not sure where to begin, here is a simple, three-step plan:

1. **Find What Feels Good:** Experiment with different types of movement. Is it walking? Dancing? Cycling? Trying out a new class? The goal isn't perfection—it's finding joy.
2. **Make It Manageable:** Start small. Just 10 minutes a day can make an enormous difference. The key is consistency, not intensity.
3. **Pair It with Joy:** Listen to music, take in nature, or celebrate small wins. Focus on how movement makes you feel, not just what it looks like.

Closing Thought: The Gift of Movement

The truth is, movement is so much more than burning calories, getting in your daily steps, or hitting a certain number on the scale. It's about feeling energized when you wake up, carrying groceries with ease, or chasing your kids or grandkids at the park without hesitation.

It's about creating a body that works with you, not against you. A body that lets you live fully, with strength, vitality, and confidence.

As you continue your journey, remember movement isn't something you "have to do." It's something you *get* to do. It is a celebration of what your body can achieve and a gift to your future self.

So take that first step, however small it may seem, and trust that every movement matters. You're not just building a stronger body—you are building a life that supports your dreams, one step, one rep, one joyful moment at a time.

You've got this.

Your Turn: Integrate & Empower

Mini Exercise: Moving in a Way That Works for You

📌 *Want to go deeper? A link to the full Whole-BEING Empowerment Workbook is available in the Resources section of this book.*

Movement is a powerful tool for your health and well-being, but it should also be something you enjoy. This exercise will help you find movement that works for your body and lifestyle while setting a plan for consistency.

Step 1: Reframe Your Mindset

Many people associate exercise with obligation or punishment, but movement is a form of self-care. Take a moment to reframe how you think about it.

- **Old Mindset:** "I have to exercise to lose weight."
- **New Mindset:** "I get to move my body in a way that makes me feel strong and energized."

What is one negative belief you have held about exercise? Rewrite it in a positive way.

Step 2: Identify What Feels Good

You don't have to follow a rigid workout plan. The best movement is the one you enjoy and can sustain.

- Think about times when movement has felt enjoyable—walking, dancing, swimming, stretching, hiking, or even playing with your kids or grandkids.
- Pick two to three types of movement that feel fun or rewarding.

Step 3: Set a Realistic Movement Goal

Start with a simple, achievable goal. Instead of aiming for perfection, focus on consistency.

- Example: "I will walk for 10 minutes after dinner three times a week."
- Example: "I will do a five-minute stretch before bed every night."

Choose one movement goal for the next week.

Step 4: Hydration Check-In

Movement and hydration go hand in hand. Staying properly hydrated supports muscle function, energy levels, and recovery.

- Are you drinking enough water before and after exercise?
- Aim for **1 ounce of water per kilogram of body weight daily**. Increase this if you are sweating more due to exercise.

Set a hydration goal:

- "I will drink a full glass of water before and after exercise."
- "I will carry a water bottle with me throughout the day to stay on track."

Step 5: Build Accountability

Sticking to movement is easier when you have a system in place. Choose one method to keep yourself accountable:

- **Track it:** Use a calendar, app, or journal to log your activity.
- **Tell someone:** Share your goal with a friend or workout friend.
- **Pair it with something you enjoy:** Listen to a podcast while walking or watch a show while stretching.
- **Create a reward system:** Set up a non-food reward for consistency, like buying new workout gear or scheduling a relaxing activity.

Continued Actions:

1. **Keep it flexible:** Life happens. If you miss a planned workout, move your body in another way that day.
2. **Check in weekly:** Are you enjoying your movement routine? Adjust as needed to keep it sustainable.
3. **Prioritize hydration:** Keep a water bottle with you, drink a glass of water first thing in the morning, and check in on your hydration throughout the day.
4. **Celebrate progress:** Every time you move, you're investing in your strength, energy, and well-being—acknowledge that win!

Final Thought: Movement isn't about obligation—it is about honoring the incredible body you have. Stay hydrated, keep moving in ways that feel good, and enjoy the process.

Chapter 11

Stress, Sleep, and Whole-BEING Wellness

Restoring Balance for a Thriving Body and Mind

Introduction: The Weight of Stress on Body and Soul

Stress doesn't just live in your mind—it takes root in your body, your habits, and your health. It changes the way you think, the way you move, the way you eat, and even the way you see yourself. I know this truth deeply because I've lived it.

In my thirties, I wore the mask of success. I had a prestigious role at a world-renowned clinical research hospital, a beautiful home inside the beltway, and a career trajectory that promised a bright future. On the surface, everything looked picture-perfect. But beneath that polished exterior, cracks were quietly forming—ones I refused to acknowledge.

Amid the busyness of my career, I could not ignore the quiet but relentless ticking of my biological clock. With each passing year, the sound grew louder, filling me with urgency and fear. I wanted to believe there was still time, but deep down, I worried it might already be slipping through my fingers.

After months of waiting, hoping, and holding my breath, a positive pregnancy test finally arrived. For a brief moment, I allowed myself to feel joy and possibility. But that hope was stolen away by miscarriage. The pain was devastating, and it wasn't a one-time event. Miscarriage after miscarriage followed, along with a failed round of In vitro fertilization (IVF). Each time, I felt like I was losing a piece of myself—my dreams, my sense of identity, my belief in the future.

At the same time, I was stuck in a toxic relationship that magnified my stress. It was an unhealthy, controlling partnership, layered with anger and blame. I couldn't see it clearly then—my desire for motherhood blinded me to the cracks. I poured everything into chasing what I thought would make me whole, while the foundation of my well-being was crumbling beneath me.

The stress was suffocating, relentless. I felt like I was treading water with weights strapped to my ankles—barely staying afloat, completely overwhelmed. It was coming from every direction: the pressure of a high-stakes career, the grief of personal loss, and the emotional turmoil of the abusive relationship.

And here is what I now understand: Stress doesn't stay in your head—it infiltrates your entire body. Chronic stress disrupts your hormones, weakens your immune system, and fuels inflammation. It robs you of sleep, energy, and clarity. For me, it showed up as endometriosis, relentless joint pain and damage, and unexplained weight gain. The physical toll became impossible to ignore. I was exhausted, not just physically but emotionally and spiritually.

If you've ever felt like stress has consumed every part of you, you're not alone. Maybe you have been using GLP-1 medications, finding relief from hunger and cravings, but the weight of stress still lingers. Maybe you're carrying more than just physical weight—it's the weight of expectations, guilt, or feeling like you're always falling short.

But here's what I learned: The body is remarkably resilient. Your body wants to heal—it's *literally designed to heal.*

When you begin to release the grip of stress and take small steps to nourish yourself—mind, body, and spirit—change is possible. It's not instant, and it is not easy. But it is worth it.

This chapter is about empowering you to reclaim your well-being, no matter where you're starting from. Together, we'll explore how stress, sleep, and resilience are deeply interconnected, and how you can create an environment where your body thrives.

You don't need to be perfect. You don't need to fix everything at once. You just need to take one step. Then another. Because growth isn't about never falling—it's about getting back up every time you do.

Let's begin.

The Link Between Stress and Weight

Stress doesn't just feel heavy—it adds weight to your body in very real ways. Chronic stress activates hormonal responses that promote fat storage, intensify cravings, and disrupt metabolic function. Here's how it works:

1. **Cortisol's Role:** Stress activates your body's fight-or-flight response, releasing cortisol, the "stress hormone." In small doses, cortisol is helpful—it gives you the energy to respond to danger. But chronic stress keeps cortisol levels elevated, leading to fat storage, especially around your abdomen.
2. **Cravings and Emotional Eating:** High cortisol levels often increase cravings for high-fat, high-sugar "comfort foods." These foods temporarily soothe your brain but leave you stuck in a cycle of stress and overconsumption.
3. **Disrupted Metabolism:** Chronic stress can lower your metabolic rate, making it harder to burn calories efficiently.

The Science Behind Stress and Inflammation: What's Happening in Your Body

As a healthcare clinician and leader with years of experience, I have seen firsthand how stress impacts the body—not just in theory, but in measurable, biological ways. Stress isn't just a feeling; it is a cascade of physiological responses that can profoundly affect your health.

When you're stressed, your body activates its sympathetic nervous system (remember fight-or-flight response?) releasing stress hormones like cortisol and adrenaline. While this response is helpful in occasional short bursts—helping you escape danger or rise to a challenge—chronic stress keeps your body in a constant state of high alert. Over time, this will bring down your immune system and creates a perfect storm for inflammation.

Research has shown that chronic stress interferes with the body's ability to regulate cortisol levels. Elevated cortisol triggers an inflammatory response, which, while essential for healing in acute situations, becomes harmful when it is persistent. This chronic inflammation has been linked to a wide range of health issues, including:

- **Arthritis**: Prolonged inflammation damages joint tissues, leading to stiffness, swelling, and pain.
- **Cardiovascular Disease**: Chronic stress increases blood pressure and promotes the buildup of arterial plaque.
- **Metabolic Disorders**: Elevated cortisol levels can lead to insulin resistance, weight gain (especially around the abdomen), and type 2 diabetes.
- **Mental Health Disorders**: High stress can exacerbate anxiety, depression, and burnout, creating a vicious cycle that affects both mind and body.

Identifying Stress in Your Life: Signs You Might Be Overwhelmed

Sometimes, stress becomes so ingrained in your daily life that you do not even notice its presence. But your body is always giving you signals—if you know what to look for.

Here are the common ways stress shows up:

1. **Physical Symptoms**: Do you frequently experience headaches, muscle tension, digestive issues, or trouble sleeping? Chronic stress can manifest physically, even when you're not consciously aware of feeling stressed.
2. **Emotional Clues**: Have you been feeling irritable, anxious, or overwhelmed? High stress often amplifies negative emotions and makes it harder to regulate your mood.
3. **Behavioral Patterns**: Do you find yourself stress-eating, skipping workouts, procrastinating, or turning to unhealthy habits like excessive screen time or alcohol? Stress often triggers behaviors that provide temporary relief but do not serve your long-term goals.
4. **Cognitive Fog**: Do you struggle to focus, make decisions, or remember things? Chronic stress can impair cognitive function, making it harder to stay sharp and productive.

Take a moment to reflect on your own life. Are any of these signs showing up for you? Recognizing them is the first step toward regaining control.

Concrete Evidence of Stress's Impact

If you're someone who appreciates measurable data, consider these findings:

- **Increased Inflammation Markers**: Studies show that people experiencing chronic stress have higher levels of C-reactive protein

(CRP), a marker of systemic inflammation associated with heart disease and other conditions.
- **Shortened Telomeres:** Telomeres, the protective caps at the ends of your DNA chromosomes, are shorter in individuals under chronic stress. This is significant because shorter telomeres are linked to accelerated aging and an increased risk of age-related diseases. As part of my work with clients, we utilize advanced testing to assess telomere length and evaluate biological versus calendar aging. These insights offer a clearer picture of how stress and other factors impact your overall health and accelerate the aging process, including the biological aging of ten distinct organ systems—such as your heart, brain, and immune system.
- **Gut Microbiome Disruption:** Chronic stress alters the balance of gut bacteria, which can lead to digestive issues, reduced immunity, and even mood disorders.
- **Sleep Disruption:** High stress can interfere with the production of melatonin, making it harder to fall and stay asleep. Poor sleep, in turn, exacerbates inflammation and reduces your body's ability to recover.

Taking Steps to Regain Control of Your Stress

The good news? Once you identify stress as a contributor to your challenges, you can begin to take proactive steps to reduce its impact. Here's how:

1. **Name It to Tame It:** Start by acknowledging the stress in your life. Sometimes, simply naming the source of stress—whether it's a demanding job, relationship issues, or health concerns—can help you feel more in control.
2. **Prioritize Recovery:** View stress management as an essential part of your health, not a luxury. Just as you wouldn't ignore a physical injury, don't ignore the toll stress takes on your body.

3. **Create a Toolkit**: Develop daily practices that reduce stress and support resilience. These might include:
   - Practicing mindfulness or meditation to calm your nervous system.
   - Engaging in light movement like yoga or walking to release tension.
   - Journaling to process your thoughts and emotions.
   - Prioritizing sleep hygiene to ensure your body has time to recover.
4. **Seek Support**: Whether it's talking to a trusted friend, joining a supportive community, or seeking professional guidance, you don't have to navigate stress alone.

This chapter will help you take meaningful steps to address stress—not just to feel better in the moment, but to protect your long-term health. By learning how to tune into your body's signals and respond with intention, you will build the resilience needed to thrive.

The Gut-Stress Connection: How Microbiome Health Affects Your Well-Being

Chronic stress doesn't just affect your mind and immune system—it also disrupts the delicate balance of your gut microbiome. Your gut is often called your "second brain" for a reason. It is home to trillions of bacteria that influence digestion, immunity, nutrient absorption, and even mood regulation. When stress becomes chronic, it can disrupt this ecosystem, leading to issues like digestive discomfort, reduced immunity, inflammation, and mood disorders like anxiety and depression.

Advanced Microbiome Testing: Insights for Personalized Wellness

As part of my approach to whole-being wellness, I offer advanced microbiome testing to give clients a complete picture of their gut health. This innovative testing provides an in-depth analysis of an individual's gut microbiome diversity and functionality, delivering insights into:

- **Short-Chain Fatty Acid Production:** Critical for gut lining health, reduced inflammation, and energy balance.
- **Amino Acid and Vitamin Production:** Essential for overall health, muscle repair, and optimal immune function.
- **Sugar Utilization:** How your gut processes sugars, which can affect cravings, energy levels, and metabolism.

One of the most valuable aspects of this testing is the actionable recommendations it provides. Based on your unique microbiome profile, you will receive personalized guidance on the specific foods and nutrients that will help rebalance your gut. Whether it's adding more fiber-rich vegetables, incorporating fermented foods, or supplementing with specific probiotics, these insights take the guesswork out of improving your gut health.

The Gut-Stress Connection

The health of your gut has far-reaching implications beyond digestion:

- **Mood Regulation:** Over 90% of serotonin—the "feel-good" neurotransmitter—is produced in the gut. A disrupted microbiome can interfere with mood balance, making stress harder to manage.
- **Immune Function:** A sizable portion of your immune system resides in the gut, so a balanced microbiome is essential for preventing illness and managing inflammation.

- **Cognitive Health:** The gut-brain axis is a direct communication line between your gut and brain, influencing focus, memory, and overall mental clarity.

Carmen's Story: A Journey to Gut Health and Empowered Weight Loss

When Carmen first came to me, she was seeking DNA-based coaching for weight loss. Like so many others, she had tried countless diets, but nothing seemed to stick. On top of that, she was battling fatigue, bloating, and brain fog—symptoms she chalked up to the stress of managing a demanding career and her busy family life. She felt depleted, out of sync with her body, and unsure where to start.

Through her DNA Blueprint, she discovered critical insights about her body's unique needs, including a predisposition to lower protein utilization and reduced metabolic flexibility. But what really caught our attention was her gut health. Using advanced microbiome testing, we uncovered a significant imbalance in her gut bacteria, including low diversity and reduced short-chain fatty acid production. These findings explained many of her symptoms, including her digestion issues, energy dips, and even her struggles with cravings.

It was clear that addressing her gut health would be foundational—not just for improving her digestion but for creating the optimal environment for weight loss and mental clarity. Together, we created a personalized plan tailored to Carmen's genetic and microbiome results.

Carmen began incorporating more fiber-rich foods, like leafy greens, oats, and berries, which supported the growth of beneficial gut bacteria. She added fermented foods like kimchi, sauerkraut, and Greek yogurt to improve her microbiome diversity. To target her specific microbiome imbalances, we included a high-quality probiotic regimen and made sure she was getting the essential nutrients her body needed for metabolic efficiency.

The results were transformative. Within just a few weeks, Carmen noticed a dramatic improvement in her digestion and energy levels. She was thinking more clearly, and her mood stabilized. By balancing her gut health, she also experienced fewer cravings and a noticeable reduction in bloating, which had been a significant barrier to feeling confident.

Over time, Carmen didn't just lose weight—she gained something far more valuable: a sense of empowerment and control over her health. Her weight loss felt sustainable for the first time in years, and her newfound vitality spilled over into every aspect of her life. Tasks that once felt overwhelming became manageable, and she started prioritizing movement and self-care in ways that worked for her lifestyle.

Carmen's journey is a powerful reminder of the interconnectedness of our bodies. When we address root causes, like gut health and genetic predispositions, we set the stage for true transformation. By taking control of her microbiome and aligning her habits with her DNA Blueprint, Carmen unlocked a healthier, more vibrant version of herself—and she has never looked back.

Why Sleep is Non-Negotiable

If stress is *the fire*, poor sleep is *the gasoline*. Without adequate rest, your body can't recover from the wear and tear of daily life. Sleep isn't just about feeling rested—it's when your body performs critical functions like repairing tissues, balancing hormones, and strengthening your immune system. It's no exaggeration to say that sleep is the foundation of health, affecting everything from your weight to your mental clarity.

But here's the challenge: in today's fast-paced world, sleep often takes a backseat to late-night work emails, Netflix marathons, or endless scrolling on social media. Over time, this lack of rest doesn't just leave you tired—it leaves your body running on empty.

How Poor Sleep Disrupts Metabolism and Weight Loss

1. **Increased Hunger Hormones:** Sleep deprivation disrupts the delicate balance of hunger-regulating hormones. When you're short on sleep, your body produces more ghrelin (the hormone that makes you feel hungry) and less leptin (the hormone that signals fullness). The result? You feel hungrier, crave high-calorie comfort foods, and find it harder to feel satisfied after meals.

2. **Reduced Willpower and Decision-Making:** Sleep loss affects the prefrontal cortex, the part of your brain responsible for decision-making and impulse control. This makes it much harder to resist cravings, choose healthier foods, or stick to your exercise routine. If you've ever reached for a sugary snack after a poor night's sleep, you're not alone—your brain is seeking a quick energy boost to make up for the lack of rest.

3. **Elevated Cortisol and Stress Hormones:** Poor sleep increases cortisol, your body's primary stress hormone. Elevated cortisol doesn't just leave you feeling anxious—remember it also encourages your body to store fat, particularly around the midsection. When this happens night after night, it can create a vicious cycle: stress disrupts sleep, and poor sleep compounds the effects of stress on your body.

4. **Genetic Influences on Sleep Quality and Duration:** Your genes also play a role in how well you sleep. Variations in certain genes, such as **PER3**, **CLOCK**, and **DEC2**, influence your natural sleep duration, your sensitivity to light, and your body's circadian rhythm (think internal clock). For example, some people are genetically predisposed to require more sleep, while others thrive on shorter durations. However, even those with a genetic predisposition to shorter sleep still need quality rest to function at their best.

Genetic testing can provide valuable insights into your sleep patterns, helping you tailor your habits to optimize your rest. By understanding your DNA, you can work with your body instead of against it, setting yourself up for deeper, more restorative sleep.

Practical Tips for Better Sleep

1. **Create a Sleep Sanctuary:** Your bedroom should be a haven for rest. Keep the room cool (around 65°F), dark (blackout curtains work wonders), and quiet (consider a white noise machine if needed). Invest in a comfortable mattress and pillows that support restful sleep.

2. **Wind Down with Intention**: Establish a calming pre-bedtime routine to signal to your body that it's time to relax. This might include gentle yoga, meditation, journaling, or reading a book. Avoid stimulating activities, like intense exercise or work emails, in the hour leading up to bedtime.

3. **Limit Screen Time**: Blue light from phones, tablets, and computers interferes with the production of melatonin, the hormone that helps regulate sleep. Try to power down devices at least an hour before bed. If that's not possible, consider using blue light-blocking glasses or enabling the "night mode" setting on your devices.

4. **Stick to a Schedule**: Consistency is key. Strive to go to bed and wake up at the same time every day, even on weekends. This helps regulate your circadian rhythm, making it easier to fall asleep and wake up feeling refreshed.

5. **Fuel Your Sleep with Nutrition**: Certain nutrients, like magnesium, zinc, and tryptophan, promote relaxation and better sleep. Foods like almonds, bananas, turkey, and leafy greens can support your body's ability to rest. On the flip side, avoid heavy meals, alcohol, and caffeine in the evening.

6. **Consider Your Genetics:** If you know your genetic predispositions through testing, you can tailor your habits to your unique biology. For example, if your DNA suggests you're more sensitive to caffeine, avoiding coffee after noon can make a significant difference in your sleep quality.

Sleep… The Ultimate Reset Button

When you prioritize sleep, everything else in your life improves—your energy, your mood, your ability to make healthy choices, and even your body's ability to lose weight and build muscle. If you've relied on GLP-1 medications in the past, you might find that prioritizing sleep becomes one of the most powerful tools in your toolkit for maintaining progress and living with vitality.

Your body is designed to heal, reset, and thrive. By giving yourself the gift of restful sleep, you're not just improving your nights—you're transforming your days.

Holistic Habits for Resilience

To thrive, you need more than stress management and sleep—you need resilience. Resilience is the ability to adapt, recover, and grow stronger in the face of challenges. It's not just about bouncing back—it's about rising higher. Cultivating habits that nurture your body, mind, and spirit is the key to unlocking this strength within yourself.

1. **Move Your Body:** Regular physical activity is one of the most powerful ways to reduce stress, lower cortisol, and elevate your mood. Movement isn't just about exercise—it's about celebrating what your body can do. Whether it's a brisk hike on a cold morning, dancing with your granddaughter, or lifting weights, every step you take is an investment in your well-being.
2. **Nourish with Intention:** Food is more than sustenance—it's information for your cells. Focus on nutrient-rich foods that

combat inflammation, like leafy greens, fatty fish, nuts, seeds, and vibrant fruits. Every bite you take is a step toward healing and strength.
3. **Connect with Others**: Humans are wired for connection. Loneliness doesn't just affect your mood—it amplifies stress and can harm your health. Build a circle of people who uplift and inspire you. Share meals, call a friend, or join a community. Remember, you don't have to walk this journey alone.
4. **Practice Gratitude**: Gratitude is a phenomenal force that shifts your focus from what's wrong to what's possible. Each day, write down three things you're grateful for, no matter how small. It could be the warmth of your morning coffee, a kind word from a friend, or the strength you showed today. Gratitude doesn't erase challenges, but it empowers you to face them with optimism and grace.

Your Journey to Whole-BEING

Stress and poor sleep can feel like an endless cycle, but here's the truth: you are not stuck. You have the power to rewrite your story, starting right now. Your body is an incredible machine, designed to heal, repair, and thrive when given the right tools and support.

Imagine waking up each day feeling energized, confident, and calm—ready to take on whatever comes your way. That's not a distant dream—it's a plan within your reach. By addressing stress, prioritizing sleep, and building resilience, you're creating a foundation for a life that's not just healthier but also unstoppable.

I've been in the trenches of stress, exhaustion, and doubt—and I've come out stronger. My journey taught me that growth isn't just possible—it's inevitable when you choose to invest in yourself.

This is your moment to take the next step toward wholeness. The tools and insights in this chapter are yours to use, and I'm here to remind you: You've got this. You are capable. You are unstoppable. Let's move forward together into the life you deserve.

Gut Microbiome and Stress Research:

1. **Cryan, J. F., Dinan, T. G. (2012).** The gut-brain axis: Pathways for communication and implications for stress-related disorders. *Frontiers in Neuroscience*, 6, 119. doi:10.3389/fnins.2012.00119
   - This study highlights the role of the gut-brain axis in stress response, noting how microbiome disruptions contribute to mood disorders, immunity, and inflammation.
2. **De Palma, G., Blennerhassett, P., Lu, J., et al. (2015).** Microbiota and host determinants of behavioural phenotype in maternally separated mice. *Nature Communications*, 6, 7735. doi:10.1038/ncomms8735
   - Research showing how chronic stress alters gut bacteria and leads to behavioral changes linked to mood regulation and cognitive function.
3. **Mayer, E. A., Knight, R., Mazmanian, S. K., et al. (2014).** Gut microbes and the brain: Paradigm shift in neuroscience. *The Journal of Neuroscience*, 34(46), 15490–15496. doi:10.1523/JNEUROSCI.3299-14.2014
   - This review explains the impact of gut microbiota on brain function, highlighting its role in mood, anxiety, and depression.

Sleep Reference:

Hirshkowitz, M., & Sharafkhaneh, A. (2019). Genetics of Sleep: Understanding How Genes Influence Sleep and Circadian Rhythms. *Sleep Medicine Clinics*, 14(3), 329–340. doi:10.1016/j.jsmc.2019.04.001

This article delves into the genetic regulation of sleep and the role of key genes, such as **PER3**, **CLOCK**, and **DEC2**, in shaping sleep patterns, duration, and circadian rhythms. These findings highlight how genetic predispositions can influence individual variations in sleep needs and quality.

Your Turn: Integrate & Empower

Mini Exercise: Stress, Sleep, and Whole-BEING Wellness

Restoring Balance for a Thriving Body and Mind

📌 *Want to go deeper? A link to the full Whole-BEING Empowerment Workbook is available in the Resources section of this book.*

Stress and sleep have a profound impact on your overall well-being. This exercise will help you identify stress triggers, develop strategies for resilience, and create a sleep routine that supports whole-body health.

Step 1: Identifying Stress Patterns

Awareness is the first step in breaking free from chronic stress. Take a moment to reflect on your daily life and answer the following:

- What are the top three sources of stress in your life right now?

    1. _____

    2. _____

    3. _____

- How does stress show up in your body? (Check all that apply)

☐ Headaches or muscle tension

☐ Digestive issues (bloating, nausea)

☐ Emotional overwhelm (irritability, anxiety, sadness)

☐ Sleep disturbances (trouble falling or staying asleep)

☐ Difficulty getting out of bed most mornings

☐ Increased cravings or emotional eating

☐ Other: _____

- What is one small action you can take today to lower stress in your life?

_____

Step 2: Your Stress-Relief Toolkit

Managing stress requires proactive habits. Choose two to three techniques you can commit to:

☐ Deep breathing exercises (e.g., box breathing)

☐ Journaling for clarity and emotional processing

☐ Engaging in physical movement (e.g., walking, yoga, strength training)

☐ Practicing mindfulness or meditation

☐ Connecting with supportive friends or family

☐ Setting boundaries with work, technology, or relationships

☐ Prioritizing self-care activities (e.g., reading, creative hobbies, taking a bath)

☐ Other: _____

Which one will you commit to practicing today?

_____

Step 3: Evaluating Your Sleep Habits

Sleep is the foundation of health. Let's assess your current sleep patterns:

How many hours of sleep do you typically get per night?

☐ Less than five hours

☐ Five to six hours

☐ Seven to eight hours

☐ More than eight hours

Do you wake up feeling rested?

☐ Yes

☐ No

Which of the following could be interfering with your sleep? (Check all that apply)

☐ Stress or racing thoughts

☐ Blue light exposure before bed (TV, phone, laptop)

☐ Irregular sleep schedule

☐ Caffeine or alcohol intake in the evening

☐ Poor sleep environment (too hot, noisy, or uncomfortable bed)

☐ Other: _____

Step 4: Creating a Sleep-Enhancing Routine

Quality sleep starts with intention. Choose two to three habits to implement tonight:

☐ Set a consistent bedtime and wake-up time

☐ Reduce screen time at least one hour before bed

☐ Create a relaxing bedtime routine (e.g., reading, stretching, meditating)

☐ Optimize your sleep environment (cool room, blackout curtains, white noise)

☐ Avoid caffeine and alcohol in the evening

☐ Incorporate magnesium-rich foods or supplements for relaxation

☐ Other: _____

Which one will you start tonight?

_____

Continued Actions for Whole-BEING Wellness

To support your resilience, aim to:

1. **Move Daily:** Build to at least 30 minutes of movement to reduce stress and improve sleep.
2. **Stay Hydrated:** Drink at least **1 oz per kg of body weight** daily for optimal hydration.
3. **Nourish Your Gut:** Support your microbiome with fiber-rich foods, fermented foods, and probiotic sources.

4. **Practice Gratitude:** Write down three things you're grateful for each night before bed.

Your body is designed to heal, but it needs the right support. Small, consistent actions create lasting change. Choose one step to take today, and trust that each choice is moving you toward a stronger, more balanced you.

Final Reflection

What is one key takeaway from this exercise that you can apply to your daily routine?

Remember: **You are in control of your well-being.** Each small step adds up to a healthier, more resilient life. Keep going—you've got this!

Part 5

Thriving for Life

Your Path to Lasting Freedom

"You didn't come this far to only come this far."

— Unknown

Owning Your Success and Sustaining It Without GLP-1s

You've done the work. You've built the habits. Now, it's time for the final step—learning how to maintain your progress for life.

Part Five will guide you through tapering off GLP-1s safely and confidently. You'll learn exactly how to manage hunger naturally, prevent metabolic slowdowns, and stay in control—without relying on medication.

But this isn't just about maintaining weight loss. It's about stepping fully into your power—becoming the kind of person who doesn't need external solutions to stay healthy.

By the end of this section, you'll have a rock-solid strategy for lifelong success—one that gives you freedom, confidence, and the certainty that *you did this—and you'll continue to thrive.*

And this journey... it was *never* about the medication. It was *always* about you.

And now, as you stand at the edge of this next chapter, you might be feeling both excited and uncertain about what life looks like beyond GLP-1s.

That's completely normal. Any major transformation comes with a mix of confidence and fear. But let me remind you of something...

You did this—not the medication.

GLP-1s may have helped quiet the noise of cravings, but every healthy choice you made—every time you nourished your body, moved with intention, reframed a setback, or proved to yourself that you are capable—that was you. The medication didn't make those decisions. You did.

And that means you already have everything you need to sustain your success.

This concluding section of the book will help you navigate the transition away from GLP-1s with confidence and clarity. We'll cover exactly how to taper off (in partnership with your healthcare provider), how to manage appetite and metabolism naturally, and how to set yourself up for a thriving, empowered future.

But before we get into the details, let's take a moment to acknowledge just how far you've come...

A Quick Look Back: The Foundation You Have Built

Part 1: Understanding the Journey – You learned how GLP-1 medications work and the role of genetics, metabolism, and insulin resistance in weight loss. You also saw that true, lasting health isn't just about a number on the scale—it's about building strength, energy, and resilience.

Part 2: Reclaiming Your Mindset & Empowerment – You rewired your thoughts, broke free from limiting beliefs, and reshaped your relationship with food. You learned how to interrupt self-sabotage, build unshakable confidence, and create a mindset that makes success inevitable.

Part 3: Personalizing Your Plan for Lasting Results – You discovered the power of DNA-based coaching, understanding how your body uniquely responds to food, exercise, and stress. If you're harnessing the power of your unique genetic makeup, you're no longer following generic advice—you're working with your body, not against it.

Part 4: Sustainable Habits for Whole-BEING Wellness – You mastered nutrition, movement, stress resilience, and sleep—the four pillars that truly determine long-term success. You learned to nourish your body, not deprive it. You found ways to move that feel good, not like punishment. And you took control of the daily habits that shaped your future.

You've likely recognized this is so much more than a "diet" or weight-loss plan.

You've transformed how you think, act, and live.

And now, it's time for the finishing step—sustaining it for life.

Chapter 12

Breaking Free—Strategies for Tapering Off GLP-1 Medications

Knowing When It's Time to Transition Off GLP-1s

"Freedom doesn't come from what you quit. It comes from what you build."

— Holli Bradish-Lane

For many, starting GLP-1 medications is a moment of hope—a breakthrough after years of struggling with weight, cravings, and metabolic health. The early weeks often bring dramatic changes: hunger fades, the scale moves, and for the first time in years, food doesn't feel like a battle.

But as the months pass, another question begins to surface:

When will I be ready to stop?

There is no universal answer, no flashing neon sign that declares: *You've arrived!* Instead, the right time to taper off GLP-1s is a deeply personal decision, influenced by both **objective markers** and **internal readiness**. Let's explore the signs that signal you might be ready—and how to make that decision with clarity and confidence.

The Science: Objective Markers of Readiness

Numbers alone don't dictate success, but they do offer valuable insights. Consider these key indicators:

- **Body Mass Index (BMI):** A BMI between **18.5 and 24.9** is considered a healthy range. If you started with obesity (BMI >30) and have reached the **overweight or healthy category**, this may be a cue to assess your next steps. However, BMI isn't everything—**body composition** (muscle vs. fat), waist-to-hip ratio, and metabolic markers are equally important.

- **Stable weight for several months:** If your weight loss has plateaued at a level where you feel comfortable, energetic, and metabolically healthy, that's a good sign. Rapid fluctuations or continued loss beyond your goal may mean your body still needs support.

- **Improved metabolic health:** Have your **blood sugar, insulin sensitivity, and inflammatory markers** improved? If your lab work shows positive changes—and you're maintaining them without extreme effort—it may be time to consider life beyond GLP-1s.

The Feeling: Subjective Signs It's Time

Beyond the data, your experience in your body matters. Take a moment to check in with yourself:

- ☑ **You trust yourself around food.** Can you sit down to a meal and eat intuitively—without relying on the medication to curb your appetite? Can you enjoy treats in moderation without fear of spiraling back into old habits?

- ☑ **Your daily habits are second nature.** Are you moving your body regularly, prioritizing whole foods, and sleeping well—not because you *have* to, but because it feels good? If your lifestyle changes feel effortless, you've built the resilience to sustain them.

- ☑ **Your body feels strong and capable.** This isn't just about weight—it's about how you *move* through life. Do you have the stamina for long walks? Can you lift things without strain? Do you wake up with energy instead of fatigue? Strength and vitality often signal that you're ready for the next chapter.

- ☑ **Your identity has shifted.** If you no longer see yourself as someone *trying to lose weight* but instead as someone who simply *lives a healthy life*, you've crossed a powerful mental threshold. This mindset shift is key to long-term success.

Sarah's Story: How She Knew It Was Time

Deciding to transition off GLP-1s is both a science and an art. While data can provide a guide, real-life experience often paints a clearer picture. Sarah's journey is a perfect example.

For Sarah, a 42-year-old healthcare executive and mother of two, food had always been a source of stress. She started semaglutide at 215 lbs., exhausted by years of dieting and emotional eating. At first, the medication was a revelation—her cravings disappeared, and for the first time in her life, she could stop eating when she was full.

Over 14 months, Sarah lost 70 pounds, eventually settling at 145. But it wasn't just the number on the scale that told her she was ready—it was how she felt:

- She no longer thought about food all the time. Hunger felt natural, not overpowering. She could enjoy meals without guilt or restriction.

- She had built a lifestyle she loved. Strength training three times a week, evening walks with her kids, and home-cooked meals were her new normal.
- Her health had transformed. Lab work confirmed that her insulin resistance had reversed, and she felt more energetic than she had in years.

With guidance, Sarah gradually tapered off her medication. A year later, she is still thriving—because she had shifted from relying on a drug to trusting in herself.

Are You Ready? Let's Find Out Together.

> *"Real freedom is being able to say: I trust my body now."*
>
> — Holli Bradish-Lane

Deciding to taper off GLP-1s isn't just about reaching a certain weight—it's about reaching a place of confidence, strength, and sustainability. And while you don't have to make this decision alone, it's important to do it with a plan.

This is where a personalized, DNA-based strategy can be your greatest tool.

Your genetic profile holds powerful insights into how your body processes food, burns fat, and responds to exercise. Some people thrive with a higher protein intake; others need more healthy fats. Some metabolize carbs efficiently, while others struggle with insulin resistance.

A one-size-fits-all approach won't cut it. You need a plan designed for you—one that accounts for your biology, lifestyle, and goals.

Let's create your taper strategy—together.

Schedule a **1:1 DNA-Based Weight Loss Strategy Session** with me, and let's determine:

- If your body is truly ready to taper off GLP-1s
- How to optimize your metabolism based on your unique genetics
- The best nutrition and exercise plan for long-term success
- Strategies to maintain weight without the medication

The transition isn't about "stopping" GLP-1s—it's about *stepping into the next phase* of your journey with clarity and control. You've already done the work. Now, let's make sure your success lasts for life.

Schedule your strategy session today:
https://IronCrucibleHealthScheduling.as.me/

Your future self will thank you.

You Did This—Not the Medication

Many people fear that once they stop taking GLP-1s, they'll lose control—that hunger will come roaring back, that old habits will resurface, that they'll regain everything they worked so hard for.

But let's take a step back.

Did the medication make your healthy choices for you? No.

Did it wake up early and go for a walk? No.

Did it cook a protein-rich meal, drink water, or choose to step away from emotional eating? No.

You did those things.

GLP-1s may have given you an initial boost, but the real transformation has been in the habits and strategies you've built. And those don't disappear just because the medication does.

That said, your body will adjust as you taper off GLP-1s, and this chapter will give you the exact tools to navigate that process with confidence and ease.

The Fear of Letting Go—And Why You Won't Lose Control

Maggie's Story: Reclaiming Confidence Beyond GLP-1s

It was a crisp autumn morning, the kind where the air feels sharp yet refreshing, carrying the scent of fallen leaves and freshly brewed coffee. Inside a cozy Starbucks on the corner of Main Street, I spotted Maggie before she saw me. She was sitting by the window, nervously stirring her latte, the froth long since dissolved. Her fingers traced the rim of the cup as she stared outside, lost in thought, watching the mist rise from the pavement as the city stirred to life.

When I slid into the chair across from her, she barely managed a smile before exhaling shakily. "I'm scared," she admitted, wrapping both hands around the warmth of her cup like it was the only thing grounding her. "What if I can't do this on my own?"

I could see it in her eyes—the fear, the doubt, the weight of everything she had worked for. Maggie had spent years in the exhausting cycle of dieting, emotional eating, and frustration. Then, GLP-1s had entered her life, silencing the hunger that had once consumed her thoughts. For the first time in years, she had felt in control. She had built new habits, learned to eat differently, and watched the scale reflect her progress.

But now, she was at a crossroads. Her prescription was ending, and the fear of losing it all was settling in.

I let a moment of silence pass, letting her words linger between us. Then I leaned forward and met her gaze.

"Maggie, listen to me. You did this—not the medication."

Her eyebrows furrowed, uncertain.

"GLP-1s didn't wake up early and go for a walk."

"GLP-1s didn't cook nourishing meals."

"GLP-1s didn't decide to break free from emotional eating."

I paused. **"YOU did those things."**

She let out a slow breath, her grip on the coffee cup loosening slightly.

"And if you built this success, you can keep it."

Maggie's eyes flickered with something new—something that hadn't been there when I first sat down.

Hope.

She swallowed hard. "So... I won't go back to how things were before?"

I smiled. "That's the beauty of this. **You're not the same person you were before. And that means you're never going back."**

Outside, the wind picked up, sending golden leaves swirling past the window. Maggie straightened in her chair, like she had just let go of something heavy.

She left that session with something she didn't have when she walked in.

Not just a plan. **Confidence.**

And that's exactly what you'll have as you taper off GLP-1s.

This isn't about "stopping" something—it's about owning your health.

The Science of Why You CAN Maintain Your Success

Let's put the fears to rest with some hard facts.

Metabolism is adaptable. Many people fear that stopping GLP-1s will "crash" their metabolism. But here's the truth: metabolism is dynamic. The key is stabilizing your body's energy balance through strength training, protein intake, and blood sugar regulation—all of which you've already learned.

Your gut hormones adjust over time. GLP-1 medications work by mimicking a natural hormone that affects hunger and digestion. When you stop taking them, your body doesn't just abandon you—your gut hormones adapt, especially when supported with fiber, protein, and healthy fats.

Muscle is your metabolic insurance. Losing weight without preserving muscle is a fast track to weight regain. But if you've been strength training and prioritizing protein (as we covered in Part Four), you've built metabolic protection. Muscle burns more calories at rest and keeps your metabolism stable.

Your brain is rewired. Remember all the work you did in Part Two? The neuro-associations you built around food, movement, and self-worth don't disappear overnight. You've created new thought patterns that support lasting success.

Science is on your side. You are not destined to fail—you are designed to adapt.

Common Fears About Stopping GLP-1s—And How to Overcome Them

Let's address the biggest concerns head-on:

"What if my hunger comes back too strong?"

You've already learned how to balance macronutrients, manage blood sugar, and slow down eating to recognize true hunger. These tools will keep you in control.

"What if I gain the weight back?"

Weight maintenance is about habits, not medication. You've built sustainable habits around food, movement, and stress resilience. Keep doing what works.

"What if I don't trust myself?"

Trust is built through evidence and action. Look at how far you've come. You have everything you need to succeed.

Your Taper Plan: A Personalized Approach

One of the most important things to remember is that there is no one-size-fits-all approach to transitioning off GLP-1s.

Some people taper off gradually, while others stop completely under medical guidance.

Some need extra support balancing hunger signals, while others feel a smooth transition.

Your DNA, metabolism, and lifestyle will all play a role in what works best for you.

That's why this next section is all about personalized strategies.

We'll cover:

- How to work with your doctor to safely taper off GLP-1s.
- How to manage hunger and cravings naturally through nutrition, movement, and mindset.
- How to use real-time awareness tools to stay in control of your eating habits.
- How to set up your environment for long-term success.

This is your moment.

Your success isn't dependent on a medication—it's built on the foundation you've created. And that foundation is strong.

Now, let's step into the next phase of your journey: lasting, unstoppable freedom.

How to Wean Off Safely (With Your Healthcare Provider)

One of the biggest mistakes people make when tapering off GLP-1s is rushing the process without proper medical guidance. Stopping suddenly can lead to sharp increases in appetite, blood sugar fluctuations, and metabolic shifts that feel overwhelming—not because your body is broken, but because it needs time to adapt.

To ensure a smooth, controlled taper, it is essential to partner with your healthcare provider to develop a personalized tapering plan that aligns with your body's needs.

Step 1: Follow a Gradual Tapering Schedule

A step-down approach allows your body to gradually recalibrate its hunger hormones (ghrelin and leptin) while maintaining blood sugar stability.

Typical Tapering Timeline:

Although your provider will determine the best schedule for you, a general tapering plan may look like this:

- **Weeks 1-2:** Reduce dosage to 75% of your current dose (e.g., if taking 1mg, drop to 0.75mg).
- **Weeks 3-4:** Decrease to 50% of the original dose.
- **Weeks 5-6:** Reduce further to 25% of the original dose.
- **Week 7 and beyond:** If well tolerated, discontinue completely.

If you are on a once-weekly injection, your provider may extend the interval between doses before reducing the dose itself.

If you are on a daily medication, the tapering schedule might involve small dosage reductions every 1-2 weeks until fully discontinued.

Monitoring During the Taper: Regular check-ins with your provider may include blood work every 2-4 weeks to monitor blood sugar stability, metabolic health, and potential hunger spikes.

Step 2: Work With Your Provider to Monitor Key Metabolic Markers

Your provider should monitor the following:

- ✅ **Fasting glucose, insulin levels, and HbA1c** to assess blood sugar stability.

- ✅ **Cortisol levels** to evaluate stress-related hormonal changes.

- ✅ **Gut microbiome health** to ensure smooth digestion and appetite regulation.

- ✅ **Muscle mass retention** to prevent metabolic slowdowns.

Step 3: Adjust Based on Your Body's Response

Some people taper off GLP-1s **easily**, while others **experience hunger fluctuations** as their body adapts. **Tracking your experience** (mood, appetite, cravings, energy) allows your **provider to make adjustments in real time**.

If appetite **spikes too fast**, a slower taper may be needed.

If **energy dips**, nutrition and strength training should be adjusted.

Some individuals may require a longer tapering period of eight to twelve weeks, depending on their metabolic history, blood sugar control, and response to appetite changes.

Why Your Provider Should Oversee the Process

Tapering off GLP-1s is not just about reducing the dose—it is about monitoring your body's response to ensure a successful transition. Your provider can:

- Monitor metabolic markers mentioned above—such as fasting glucose, insulin levels, and HbA1c to ensure blood sugar stability.
- Assess appetite regulation and provide dietary recommendations to prevent excessive hunger spikes.

- Identify side effects such as nausea, blood sugar crashes, or digestive changes that might require adjustments.
- Provide medication alternatives if needed, particularly for individuals managing type 2 diabetes.

Regular check-ins with your provider—whether biweekly or monthly—allow for real-time adjustments based on how your body is responding.

One of the primary ways GLP-1s helps regulate appetite is by balancing blood sugar levels— preventing energy crashes that often trigger cravings. As you taper, you will take over that role through nutrition strategies that stabilize glucose and insulin levels.

To stabilize glucose and insulin levels, focus on these three essential components of each meal, To include:

- **Protein:** Helps curb hunger by slowing digestion and maintaining lean muscle mass. Aim for 20 to 30 grams per meal from sources such as chicken, fish, eggs, Greek yogurt, or plant-based proteins.
- **Fiber:** Provides lasting fullness and supports gut health. Incorporate fiber-rich foods such as leafy greens, beans, berries, and whole grains. Aim for 30 grams/day.
- **Healthy fats:** Stabilize energy and reduce cravings. Choose avocado, nuts, seeds, and olive oil to promote satiety (feelings of fullness).

Your provider may recommend using a continuous glucose monitor (CGM) or fingerstick glucose testing during the taper can provide real-time data to identify blood sugar fluctuations and adjust meals accordingly.

Understand Potential Withdrawal Symptoms

As your body adjusts, you may experience shifts in appetite, digestion, and energy levels. These changes are temporary but important to recognize.

- Increased appetite: Hunger may return gradually or suddenly. Focus on balanced meals and mindful eating strategies covered in Chapter 13.
- Digestive changes: Some individuals experience nausea, bloating, or slower digestion as the gut readjusts. Support digestion with probiotic-rich foods and hydration.
- Blood sugar dips: Feelings of shakiness, irritability, or fatigue may indicate a blood sugar crash. Combat this by pairing carbohydrates with protein and healthy fats, such as an apple with almond butter.
- Emotional shifts: Appetite suppression from the medication may have reduced emotional eating cues. Now is the time to apply mindfulness techniques and reframe stress responses.

Tracking these changes in a journal or app can help you and your provider fine-tune your taper plan.

Adjust Based on Your Body's Response

No two people experience the taper the same way. Some find it easy, while others need more time to adjust. The key is staying flexible and being patient with yourself.

- If your appetite returns aggressively, your provider may recommend a slower taper or additional nutritional support.
- If energy levels dip, ensure you are eating enough protein and complex carbohydrates to sustain metabolism.
- If weight fluctuates, remember that temporary fluid shifts are normal and do not equate to fat gain.

Above all, this transition is not about losing something—it is about stepping fully into ownership of your health.

You built the habits. You made the changes. The medication helped, but it was not the driver of your success—you were.

As you move forward, embrace this transition with confidence and control.

💡 **Bonus Resource:** *Comprehensive Plan to Wean Off GLP-1s for Weight Loss*

🔗 *www.ironcruciblehealth.com/glp1-taper-plan*

Up next, we will explore how to manage hunger and cravings naturally, so you stay empowered in this next phase of your journey.

Managing Hunger and Cravings Naturally

One of the biggest fears people have when tapering off GLP-1s is: *What if my hunger comes back, and I lose control?*

Let's address that head-on.

Hunger isn't the enemy—it's simply a signal from your body, much like thirst or fatigue. The difference now is that you have the tools to listen and respond rather than react impulsively. You are not powerless. You have already built habits that support balance, and now you'll refine them to maintain control and confidence.

1. Recognize the Difference Between True Hunger and False Urges

Not all hunger is physical. Sometimes, we eat because of habit, emotion, or external triggers rather than actual need. Before reaching for food, take a moment to check in with yourself:

- **Am I physically hungry, or am I eating because I'm stressed, bored, or triggered by something external?**
- **Does my body actually need fuel, or am I craving food because I saw, smelled, or thought about it?**
- **Will this food truly nourish me, or am I using it as a distraction or comfort?**

By pausing before you eat, you give yourself a chance to respond with intention rather than fall into *autopilot eating*.

The Hunger Scale

Using a simple hunger scale can help you identify when to eat and when to pause.

So, you get the urge to eat.

Before jumping in, step back and rate your hunger on a scale from **1 to 10**:

- **1-2** → *Starving* – Extreme hunger, dizziness, low energy. Waiting too long to eat may lead to overeating.
- **3-4** → *Hungry* – Stomach growling, feeling ready for a meal. This is an ideal time to eat.
- **5-6** → *Neutral* – Not hungry but not full. A good place to check if you're eating out of habit or real need.
- **7-8** → *Satisfied* – Comfortable fullness, no longer hungry. Stop eating here to avoid feeling sluggish.
- **9-10** → *Overfull* – Feeling stuffed, sluggish, or regretful. Often the result of mindless or emotional eating.

Before eating, step back and rate your hunger on a scale from 1 to 10.

The goal is to start eating when you're **at a level three or four** on the hunger scale, and stop when you're **at a seven or eight**—before you reach discomfort.

Tuning into your body's signals is one of the most powerful ways to maintain control, trust yourself, and break free from old patterns.

2. Prioritize Key Components of a Satiating Meal: Protein, Fiber, and Healthy Fats

The structure of your meals will play a significant role in keeping hunger and cravings in check. When you fuel your body properly, you reduce the likelihood of intense hunger spikes or sudden cravings.

The key is balance:

- **Protein**: This is the most important macronutrient for appetite control. It slows digestion, stabilizes blood sugar, and helps preserve muscle mass. Aim to include a protein source at every meal, such as eggs, fish, chicken, Greek yogurt, tofu, or legumes. Aim for **1.2-1.5g of protein per kg of body weight, daily.**
- **Healthy Fats**: Fat doesn't just add flavor—it helps keep hunger steady and prevents the rapid blood sugar fluctuations that can lead to cravings. Incorporate sources like avocado, nuts, seeds, olive oil, and fatty fish.
- **Fiber**: Fiber slows digestion, promotes fullness, and supports gut health. Load up on non-starchy vegetables, berries, beans, and whole grains to help keep hunger in check.
- **Use Targeted Micronutrients to Support Blood Sugar Stability**: Magnesium & Zinc: Found in pumpkin seeds, spinach, cashews; Polyphenol-Rich Foods: Dark chocolate, blueberries, olive oil, green tea; Resistant Starches: Green bananas, oats, cooked-and-cooled potatoes.

3. Use Movement to Regulate Appetite

Exercise is not just about burning calories—it plays a powerful role in regulating hunger hormones and stabilizing energy levels. Movement helps balance blood sugar, reduce stress-driven cravings, and improve your overall metabolic health.

- **Strength training:** Helps preserve muscle, which supports a steady metabolism and reduces excessive hunger. Muscle tissue also increases calorie burn at rest, making it easier to maintain your progress.
- **Walking and gentle cardio:** Lowers stress hormones that often drive emotional eating and cravings. A simple walk after meals can aid digestion, help regulate blood sugar, and prevent energy crashes.
- **Higher-intensity exercise:** While appetite may temporarily increase post-workout, regular physical activity helps regulate hunger signals over time, improves insulin sensitivity, and supports long-term metabolic health.

Personalized Nutrition and Movement with Your DNA Blueprint

If you have your **DNA Blueprint**, you can take your individual plan a step further by tailoring your movement and nutrition to your unique genetic profile. Your **macronutrient ratios** (carbohydrates, proteins, and fats) play a crucial role in hunger regulation, and your genetic insights can help you determine:

- Whether your body thrives on a higher protein intake to control cravings.
- If your metabolism responds best to endurance-based or strength-focused workouts.
- How efficiently your body processes carbohydrates and fats for sustained energy.

- Whether you need additional recovery time between workouts to prevent overtraining and excessive hunger.

The goal isn't to "fight" hunger—it's to **work with your body rather than against it.** When you understand how to nourish yourself properly, move in ways that support you, and recognize **true hunger from emotional urges**, you take back control.

This transition is not about deprivation. It's about learning to trust yourself again. **And you can.**

Consider a Holistic Tool for Supporting Your Taper Naturally

As you step into this next phase, your body is adjusting to life without GLP-1 medications. While you've built the habits, mindset, and metabolic resilience to maintain your success, you may find that additional nutritional support helps smooth the taper.

Certain science-backed ingredients have been shown to naturally support appetite regulation, blood sugar balance, and gut health—helping your body continue producing its own GLP-1 efficiently.

A well-formulated approach can:

✓ Help regulate hunger and cravings naturally.

✓ Support metabolic health and energy balance.

✓ Optimize gut health, which plays a key role in appetite and digestion.

One ingredient that has gained attention in the world of metabolic health is berberine—sometimes referred to as "Nature's Ozempic."

How It Works:

- Stimulates natural GLP-1 secretion in the gut, helping to regulate appetite.
- Improves insulin sensitivity, reducing blood sugar spikes that can trigger cravings.
- Decreases cravings for carbohydrates and sugar by balancing glucose metabolism.

In my coaching program, I find real value in a natural GLP-1 support plan that aligns with these principles, using targeted nutrition and supplementation to help sustain your progress. This is not a magic pill, but rather a science-based tool designed to complement the habits you've already built.

If you're interested in additional support tailored to your journey, you can order this tool directly through this designated page: *www.ironcruciblehealth.com/natural-glp*

Having the right tools—along with the right strategies—can make all the difference in ensuring a smooth and confident taper.

Optimizing Your Environment and Awareness for Long-Term Success

I can't emphasize it enough...this journey isn't about willpower. The people who succeed long term don't have more discipline—they have better systems.

1. Set Up Your Environment for Success

- Keep your kitchen stocked with nourishing foods.
- Use smaller plates and bowls to naturally reduce portion sizes.
- Keep tempting foods out of sight—you control what's in your space.

2. Tune Into Your Eating Cues

- Use a hunger journal for a few weeks to track your appetite.
- Identify emotional eating triggers and create alternatives (walks, deep breathing, journaling).
- Try the Tomorrow Technique—YES, I am mentioning it again—if you still want it tomorrow, you can have it. (Spoiler: You often won't.)

3. Stay Connected and Accountable

- Check in with a coach or friend about your progress.
- Join a supportive community that reinforces your success.
- Celebrate wins—no matter how small.
- Use your urge management FREEDOM Toolkit (remember?)

Freedom Toolkit: Urge Management Tools

Urge Crunchers
Fast-acting physical or mental actions that redirect your urge before it takes over.
"Change your state, and you change the urge."

Pattern Interrupts
Break the loop by doing something unexpected to short-circuit the craving cycle.
"Surprise the system—disrupt the urge before it finishes its story."

The Tomorrow Technique
Use intelligent procrastination—delay the action, not deny it.
"Tell your brain; not now, maybe tomorrow."
This softens resistance and gives space for the urge to pass without shame or rebellion.

The Scramble Technique
Rewire old associations by mentally scrambling the craving's trigger image, sound, or emotion.
"Distort the urge's power by changing the picture in your mind."
Imagine the food melting, turning gray, covered in dirt, being dipped in vinegar—until it loses emotional charge.

The Freedom Toolkit: proven methods to manage the urges.

You Are Already Free: Owning Your Transformation

Maggie came back to see me three months after she had completely tapered off GLP-1s.

She looked different—not just lighter, but stronger.

"I thought I was going to fail," she admitted, "but I didn't. And now, *I actually trust myself.*"

That's the goal.

This isn't just about maintaining weight loss. It's about:

- Trusting yourself around food
- Feeling in control of your habits
- Knowing that you are fully capable of lasting success

The medication didn't create this transformation—you did.

And now?

You own it.

This is your moment. You are no longer dependent on anything outside of yourself.

From this point forward—you are free.

What Happens Next?

In the next chapter, we'll focus on long-term weight sustainability—and walk through the exact steps to navigate plateaus, overcome setbacks, and build lasting success.

Your Turn: Integrate & Empower

Mini Exercise: Breaking Free—Strategies for Tapering Off GLP-1s

Taking Control of Your Health with Confidence

Tapering off GLP-1 medications is not about losing a tool—it's about stepping fully into the power you've already built. This exercise will help you prepare for the process, create a strategy that aligns with your body's needs, and develop confidence in your ability to sustain your success.

Step 1: Reframing Your Mindset

Success is built on habits, not medication. You've already done the work—now, it's about reinforcing what you've learned.

- **What are three healthy habits you have developed while on GLP-1s that you want to continue?**

    1. _____

    2. _____

    3. _____

    What fears, if any, do you have about tapering off GLP-1s?

_____

- **How can you shift your mindset from fear of losing progress to confidence in your ability to maintain it?**

_____

Step 2: Partnering with Your Provider for a Smooth Taper

Gradual tapering, medical monitoring, and real-time adjustments will help your body adapt successfully.

- **Do you have a plan in place with your healthcare provider for tapering off your medication?**

☐ Yes, I've discussed it with my provider

☐ No, but I will schedule a conversation this week

- **Key markers to track during your taper:** (Check the ones you will monitor)

☐ Blood sugar levels (fasting glucose, HbA1c)

☐ Appetite and cravings

☐ Energy levels and mood

☐ Sleep quality

☐ Muscle mass and body composition

- **What will be your first action step to ensure a smooth taper?**

Step 3: Managing Hunger & Cravings Naturally

Your body will adapt, and you already have the tools to regulate appetite naturally.

- **Which of the following strategies will you implement to manage hunger effectively?** (Choose at least two)

☐ Eat protein with every meal to stay full longer

☐ Increase fiber intake for sustained satiety

☐ Use a hunger scale to differentiate real hunger from emotional eating

☐ Stay hydrated to reduce false hunger cues

☐ Schedule balanced meals to prevent erratic eating patterns

- **What is one meal adjustment you can make today to improve satiety and blood sugar stability?**

Step 4: Staying Consistent with Movement

Exercise is one of the most powerful ways to regulate appetite, stabilize metabolism, and maintain long-term success.

- **Which type of movement will you commit to in the next week?**

☐ Strength training (supports metabolism and lean muscle retention)

☐ Walking (helps regulate blood sugar and reduce cravings)

☐ Yoga or stretching (reduces stress and promotes mindfulness)

☐ High-intensity exercise (boosts metabolic function)

- **How many minutes of movement will you commit to each day?**

Step 5: Optimizing Your Environment for Success

Long-term success is about creating an environment that makes healthy choices easy.

- **What is one change you can make in your kitchen or home to support your success?**

☐ Stock up on whole, unprocessed foods

☐ Keep protein-rich snacks easily accessible

☐ Reduce the visibility of processed or trigger foods

☐ Prep healthy meals in advance for busy days

- **What non-food reward can you use to celebrate small victories instead of turning to food?**

Step 6: The Freedom Mindset—Trusting Yourself Again

True success is not about controlling everything perfectly—it's about trusting yourself and making empowered choices.

- **What is one piece of evidence from your journey that proves you are capable of maintaining your success?**

- **How will you remind yourself daily that YOU—not the medication—created this transformation?**

Continued Actions for Long-Term Success

- **Stay Active:** Engage in **30 minutes of movement daily** to support appetite regulation and metabolism.
- **Hydrate Well:** Drink at least **1 oz per kg of body weight** daily to support digestion and metabolic balance.
- **Maintain Protein Intake:** Aim for **1.2-1.5g of protein per kg of body weight** daily to preserve muscle and satiety.
- **Practice Mindful Eating:** Use the hunger scale to identify **true hunger vs. emotional cravings.**
- **Track Progress Without Obsession:** Use non-scale victories (energy levels, strength, confidence) as success markers.

Your transformation wasn't about the medication—it was about **YOU**. You have the knowledge, the habits, and the ability to maintain your success for life.

📌 **Want to go deeper? A link to the full *Whole-BEING Empowerment Workbook* is available in the Resources section of this book.**

Chapter 13

Your Long-Term Plan— Thriving Without GLP-1 Medications

How to Sustain Your Progress for Life

"Beware; for I am fearless, and therefore powerful."

— Frankenstein

Just think about this for a moment. You've done what once felt impossible. You've transformed your relationship with food, built new habits, and proven to yourself that you are capable of lasting change.

Now, the question is: *How do you keep this momentum going for life?*

Long-term success isn't about perfection. (Heard this a time or two before?) I'm smiling here.

It's about having a strategy for when life happens—because it will. There will be stressful days, social events, unexpected cravings, and fluctuations on the scale. The key is knowing how to navigate them without slipping back into old patterns.

This chapter will focus on the tools you need to maintain your results. You will learn how to set realistic, long-term goals that keep you motivated.

You will develop a simple, four-step process to recover quickly from any slip-ups without guilt or shame. You will gain strategies to break through weight-loss plateaus without frustration. Most importantly, you will strengthen the mindset shifts that allow you to trust your body again and know, without a doubt, that you are in control.

It is important to remind yourself that you were never dependent on GLP-1s. If you were using them, these medications may have given you a jump-start, but every healthy choice you made was yours. You were the one who jumped in and made the courageous decision to act. Took the shot (or pills). Nourished your body with intention, showed up for every workout, and reshaped your habits from the ground up. This transformation wasn't handed to you—you built it, one choice at a time.

The progress you achieved was because of your commitment, and that means you have everything you need to sustain it.

Now, it is time to step fully into this next phase of your journey—a life where food no longer controls you, where you feel strong and confident in your health, and where you know, without hesitation, that this transformation is permanent.

Let's begin.

Setting Realistic Goals for Lifelong Health

When it comes to lasting change, one of the most powerful tools you have is goal setting. But not just any goals—realistic, meaningful, and sustainable goals that align with the life you want to create.

Think back to when you first began this journey. Perhaps your initial goal was simply to lose weight or fit into a smaller clothing size. While those goals can be motivating, they are limited in scope. True transformation happens when you expand your vision beyond the scale—when you set goals that challenge you to grow in all areas of your health, strength, and resilience.

The SMART Goal Framework for LIFELONG Success

Earlier in this book, we introduced the SMART goal framework—Specific, Measurable, Achievable, Relevant, and Time-bound. You've already used this process to reach key milestones in your journey. Now, it's time to use it again to keep growing, evolving, and challenging yourself in new ways.

Setting SMART goals isn't something you do once and forget—it's a lifelong tool for continuous progress. Each phase of your journey brings new challenges and opportunities, and creating new SMART goals keeps you engaged, motivated, and striving toward your next level of success.

Robert's Story: From Weight Loss to Unstoppable Potential

When Robert first came to me, his goal was simple: lose weight and regain control of his health. He had struggled for years, trying every diet and exercise plan imaginable, only to find himself stuck in the same frustrating cycle. When he joined the DNA-based weight program at Iron Crucible Health, something finally clicked.

Within three months, Robert had lost 30 pounds—not just shedding weight but reclaiming his energy, confidence, and sense of purpose. For the first time in years, he felt like himself again.

But something unexpected happened. As he reached his goal weight, he realized that this journey wasn't just about reaching his ideal weight—it was about unlocking his full potential.

The weight loss sparked something deeper. With renewed energy and a stronger body, Robert started setting goals he once thought were impossible. Running had always been an afterthought, something he did reluctantly, out of obligation. But now, he saw it as a challenge—an opportunity to push his limits.

Post-pandemic, he reached out to me again, not because he needed to lose weight, but because he was hungry for more. He had a new goal: to become faster, stronger, and more resilient.

Over the next several months, Robert fine-tuned his training, optimized his nutrition, and pushed past barriers he never imagined he could break. And the results? Nothing short of astonishing.

He didn't just maintain his weight loss—he cut his mile run pace by an incredible two minutes.

What started as a goal to drop weight evolved into something far greater: a transformation of his body, mind, and belief in what was possible.

Robert's story is a reminder that hitting one goal is never the finish line—it's the starting point for something even bigger.

As you stand at this moment in your journey, ask yourself: *What's next?*

You've already shown that you can reshape your habits, improve your health, and take charge of your future. Now, it is time to dream bigger. To set goals that excite you. To challenge yourself in ways you never thought possible.

Because just like Robert, your journey doesn't end here. It's only the beginning.

Recovering from a Slip-Up: The Four-Step Process

Journeys are not linear. Slip-ups are not failures; they are learning opportunities. One meal, one day, or even one week off track does not erase all the progress you've made. What matters most is how you respond. The quicker you shift from self-judgment to intentional action, the easier it is to stay on course.

Let me tell you about Rachel, a client who came to me after struggling with emotional eating. One evening, after a long, stressful day, she found herself sitting on the couch with an open bag of cookies—something she hadn't done in months. By the time she realized it, half the bag was gone. The guilt hit her hard. "I just ruined everything," she thought.

But here's what I told Rachel: *One moment doesn't undo all the work you've done. What you do next is what matters most.*

If you've had a slip-up—whether it's a single indulgence or a rough few days—use this four-step process to recover quickly and regain control.

Step 1: Forgive Yourself

Everyone makes mistakes. The worst thing you can do is let guilt spiral into shame and self-sabotage.

Instead of thinking, *I messed up, so I might as well give up,* remind yourself: *This is just one moment in my journey. I am still in control.*

Take a deep breath, release the judgment, and move forward.

Step 2: Investigate the Situation

Rather than beating yourself up, get curious. What led to the slip-up?

- Were you stressed, exhausted, or overwhelmed?
- Did you skip meals and become overly hungry?
- Were you in a social setting where food was a central focus?
- Did an emotional trigger (boredom, sadness, loneliness) play a role?

Understanding the root cause allows you to prepare for similar situations in the future. In Rachel's case, she realized she had skipped lunch, then worked late, and by the time she got home, she was both starving and emotionally drained. The cookies weren't really the problem—her exhaustion and hunger were.

Step 3: Plan for Next Time

Now that you know what triggered the slip-up, create a strategy to handle it differently in the future.

If you overeat when you're stressed, plan a new response—go for a walk, call a friend, or practice deep breathing. If you get overly hungry and make impulsive choices, make sure you always have protein-rich snacks on hand. If social events throw you off, set an intention beforehand and have a plan in place (eating a balanced meal before going, drinking plenty of water, or focusing on conversations rather than food).

Planning ahead turns potential obstacles into manageable moments.

Step 4: Mentally Rehearse and Practice the Plan

Visualization is a powerful tool. Take a moment to imagine yourself in a comparable situation but managing it differently.

See yourself recognizing the trigger before it takes over. Find yourself making a choice that aligns with your goals. Mentally rehearsing this new response strengthens the neural pathways in your brain, making it more likely that you'll follow through when the moment actually happens.

Rachel used this technique before a stressful workweek. She visualized herself choosing a balanced dinner instead of reaching for cookies. When the moment came, her brain already knew the path forward.

Remember: You Are Not Starting Over—You Are Moving Forward

One setback does not erase weeks, months, or years of progress. Your success is not defined by a single choice but by your ability to recover, learn, and keep going.

The difference between those who succeed long-term and those who don't isn't *never* making mistakes—it's knowing how to move past them.

You have the tools. You have the mindset. And now, you have the plan.

The next time life throws you a challenge, you won't fall back into old patterns—you'll rise with new strength.

How to Handle Setbacks and Plateaus with Confidence

Setbacks and plateaus are an inevitable part of any long-term journey. They don't mean you're failing—they mean you're evolving. Your body is constantly adapting, and sometimes, progress isn't linear. There will be moments when the scale doesn't budge, when motivation feels out of reach, or when old habits try to creep back in. But here's what I want you to remember:

This is not the end of your progress. It's a sign that your body and mind are recalibrating for the next phase of success.

Let's break down how to navigate these moments with confidence, resilience, and renewed focus.

Reassessing Goals: Are The OLD Ones Still Serving You?

Sometimes, what worked in the beginning isn't what will take you to the next level. If you're feeling stuck, ask yourself:

- Are my goals still realistic and aligned with my lifestyle?
- Have I been too rigid, or do I need to refine my approach?
- Am I focusing too much on external measures (like the scale) and ignoring deeper signs of success?
- Is it time to write a *Dear Future Me* Letter?

Dear Future Me,

Today, I make a promise.
Not to perfection.
Not to comparison.
But to progress.

I'm choosing resilience over retreat.
Strength over stagnation.
Self-respect over self-sabotage.

I commit to honoring my body with movement,
Fueling my mind with intention,
And showing up—for me—each day.

This journey may not always be easy,
But I know who I'm becoming is worth effort.
This letter is a reminder:
Growth takes grit.
And I have it.

My Commitments

I will move my body by: _____
I will nourish myself with: _____
I will rest and recover through: _____
I will stay accountable by: _____
I will celebrate progress by: _____

Dear 'Future Me' Letter: Bonus points for prominent placement where you see it multiple times daily!

Your goals should evolve with you. If weight loss was your initial focus, perhaps now it's time to shift toward strength, endurance, or body composition improvements. Maybe your new goal is to feel energized throughout the day, improve digestion, or wake up every morning feeling well-rested and ready to tackle the day.

Progress isn't just about measuring numbers—it's about creating a life where you *feel* your best.

Changing Up Your Routine: How to Challenge Your Body in New Ways

When progress slows, it may be time to introduce variety. Your body adapts over time, and what was once effective may no longer be as challenging. If your movement or workouts feel routine or uninspiring, consider:

- Trying a new type of movement—strength training, yoga, cycling, swimming, or hiking.
- Increasing intensity with progressive overload—lifting heavier, adding resistance, or adjusting reps and sets.
- Focusing on recovery—if you've been pushing too hard, your body may need rest to reset and rebuild.

Exercise should never feel like punishment. Find movement that feels good, excites you, and keeps you engaged in the process.

Revisiting Nutrition: Small Adjustments for Big Impact

If you've hit a plateau, take a closer look at your nutrition—*not from a place of restriction, but from a place of refinement.*

- Are you getting enough protein to support muscle maintenance and satiety?
- Have portion sizes subtly increased over time?
- Is meal timing supporting stable energy levels, or are you experiencing crashes and cravings?
- Is hydration where it needs to be? Even mild dehydration can impact metabolism, digestion, and hunger cues.

If you have your DNA Blueprint, use it as a guide to fine-tune your macronutrient ratios. Some individuals thrive with more protein, while others may need to adjust carbohydrate intake based on insulin sensitivity and metabolism. Your body has unique needs—*listen to it.*

Reflecting on Non-Scale Victories: Success Beyond the Numbers

If the scale isn't moving, it doesn't mean progress isn't happening. There are countless ways to measure success that have nothing to do with weight:

- Do you feel stronger and more capable in your workouts?
- Is your digestion improving?
- Are your energy levels more stable throughout the day?
- Do you feel more confident in your skin, regardless of the number on the scale?
- Are you experiencing fewer cravings or emotional eating episodes?

Weight is just one data point. True success is about how you *feel* in your body, the habits you've built, and the strength you've gained—physically, mentally, and emotionally.

Final Thoughts: Plateaus Are Not Permanent—They're Growth Phases

A plateau is not a stop sign—it's a rest stop on the way to your destination.

Your body is not betraying you. It is adjusting, recalibrating, and preparing for what's next.

Instead of seeing a plateau as a setback, see it as an opportunity to reflect, refine, and recommit. Your progress has never been about perfection—it has always been about perseverance. And you have already proven that you are capable of overcoming obstacles.

So, take a deep breath. You are still moving forward. And when you look back, you'll realize that every challenge, every slow moment, and every adjustment was leading you exactly where you were meant to go.

Reclaiming Confidence in Your Body

One of the greatest gifts of this journey isn't just weight loss—it's the renewed confidence in your body's capabilities. You've proven to yourself that you can create change, nurture your health, and break free from the grip of cravings and old habits. But beyond that, you've learned something far more valuable: *you can trust yourself.*

For so long, you may have felt disconnected from your body—frustrated by its signals, confused by its needs, or even betrayed by its resistance to change. But look at where you are now. You've reestablished that connection. You're no longer at war with your body; you're working with it. Listening to it. Honoring it.

And that is true freedom.

Thriving Beyond the Scale

As you continue forward, I encourage you to shift your focus away from just numbers—whether that's your weight, clothing size, or calorie intake—and instead, pay attention to something deeper.

- **How do you feel when you wake up in the morning?** Do you have the energy to take on the day?
- **How strong do you feel?** Can you lift things with ease, move freely, and do the activities you love?
- **How does food make you feel?** Do your meals leave you feeling nourished and satisfied rather than controlled or restricted?
- **How do you carry yourself in the world?** Are you standing taller, speaking more confidently, and feeling more comfortable in your own skin?

True success isn't found in a single moment of achievement—it's found in the *day-to-day freedom* of knowing that you're in control of your choices, that you respect your body, and that you're capable of maintaining what you've built.

The Journey of Becoming: What This Process Has Taught You

Growth is never a single moment—it's an ongoing process of becoming.

I know this because I've lived it. Writing this book was one of the bigger challenges I've taken on. I wasn't a best-selling author, and I had my doubts about whether I could bring my vision to life. But I reminded myself of the same thing I tell my clients: you don't have to be perfect to start. You just have to begin.

That's what this journey is about—showing up for yourself, even when it feels uncertain. Stepping into the unknown, trusting that you have what it takes, and proving to yourself, repeatedly, that you are capable of more than you ever imagined.

And now, here you are. Standing at the threshold of your next chapter, equipped with the tools, strategies, and confidence to create a life that feels strong, healthy, and free.

Keep moving forward. Because you're not just reaching a goal—you're becoming the person who lives it every day.

Your Future is Yours to Shape

"Thriving is what happens when confidence meets consistency."

— Holli Bradish-Lane

Thriving isn't about reaching some perfect, final destination. It's about *creating a lifestyle* where you feel empowered, balanced, and alive. It's about waking up every morning and knowing that *you are in control*.

Your journey is not over—it's only just beginning.

You have done the work. You have faced the challenges. And now, you have *everything* you need to continue building a life that honors who you are and the vibrant future you deserve.

So, take a deep breath.

Step forward with confidence.

Just like you've stepped outside your comfort zone, I've done the same. Authoring this book wasn't easy, but I knew my purpose was to help others—just like you are learning to help yourself. Growth never stops, and neither do you.

Your best self is waiting to be fully realized. And you?

You are ready.

Your Turn: Integrate & Empower

Mini Exercise: Your Long-Term Plan for Lasting Success

You've built the habits, strengthened your mindset, and proven that lasting health is possible. Now, let's reinforce your long-term strategy and ensure your progress becomes a permanent part of your life.

Step 1: Set Your Next Goal

Now that you've mastered key habits, what's one new goal for the next three months?

(Examples: Increase strength, improve energy, master meal planning, reduce stress, build muscle)

Step 2: Plan for Challenges

Slip-ups happen. The key is knowing how to recover.

What tends to throw you off track? (Check one or more.)

- [ ] Stress/emotions
- [ ] Skipping meals
- [ ] Social events
- [ ] Fatigue/lack of sleep
- [ ] Other: _____

What is **one strategy** you can use to stay on course?

Step 3: Break Through Plateaus

If progress slows, which area might need a small tweak?

- [ ] Nutrition – More protein, fiber, or healthy fats?
- [ ] Movement – Challenging my body in new ways?
- [ ] Sleep – Getting at least 7 hours per night?
- [ ] Stress – Practicing mindfulness or movement?
- [ ] Hydration – Drinking 1 oz per kg of body weight?

What's one small adjustment you can make this week?

Step 4: Strengthen Your Confidence

What is **one habit** you've built that makes you feel proud?

What is one promise you're making to yourself moving forward?

_____ _____ _____

Final Thought

This is your life, and your transformation is yours to keep. Every choice you make is a step toward lifelong success. Keep going—because you've already proven you can.

📌 ***Want to go deeper?*** *A link to the full* **Whole-BEING Empowerment Workbook** *is available in the* **Resources** *section of this book.*

Conclusion

Empowered for Life

The Journey Continues—And You're Just Getting Started

Take a moment. Breathe in deeply. Let it sink in—you did this.

You've taken one of the most important journeys of your life. You've challenged old beliefs, rebuilt your relationship with food and movement, and proven to yourself that lasting transformation is possible. You've equipped yourself with powerful tools—mindset shifts, DNA-based insights, and sustainable habits that will carry you forward long after this book ends.

But this isn't the end. It's the beginning.

Celebrate Your Progress

"The best view comes after the hardest climb."

— Unknown

Think back to where you started. The fears, the doubts, the frustration of trying everything and wondering if this time would be any different.

And now look at you. You've done the work.

You've broken free from quick fixes and temporary solutions. You've reclaimed control over your body, your health, and your future.

Pause for a moment and acknowledge your own success. Not just the number on the scale, but the real victories:

- The energy you've gained.
- The confidence in knowing you can trust yourself.
- The strength you've built—inside and out.
- The freedom from emotional eating, cravings, and the feeling of being stuck.

No one handed you this transformation—you *created* it.

The Power of What You Have Learned

Success isn't about luck or willpower—it's about having the right tools. And now, you have them.

- **Mindset:** You've rewired your beliefs and broken free from self-sabotage. You see challenges as opportunities for growth rather than roadblocks.
- **DNA-Based Insights:** You understand *your* body—not someone else's, not a one-size-fits-all plan, but a personalized roadmap designed for your unique biology.
- **Sustainable Habits:** You're no longer chasing quick fixes. You've built daily rituals that support your health and make success inevitable.
- **Certainty:** You don't have to *hope* you'll maintain your progress. You *know* you will, because you've built a foundation that lasts.

This knowledge is yours forever. No matter where life takes you, you can always return to the tools and strategies you've learned here in this guide.

Your Vital Journey is Just Beginning

Transformation doesn't stop when you reach a goal—it evolves with you. The more you grow, the more you realize how much you're capable of.

That's been my experience too. Drafting this book was one of the biggest leaps of faith I've ever taken. I've always had a creative side and a passion for science, health, and coaching. But taking my personal journey, professional insights, and years of experience and putting them into a book for the world to see? That was daunting.

For a long time, I questioned if I was "qualified" to write a book. I wasn't a best-selling author or a polished writer—I was someone who had lived the struggles of weight loss, seen firsthand how our healthcare system often fails people, and spent years coaching individuals to reclaim their health. But I realized something important: just like you don't need to be perfect to start your transformation, I didn't need to be a perfect writer to share something meaningful.

So, I sat down, faced my doubts, and put my heart into these pages. I authored this book not to impress but to *empower*—to give you every tool, every strategy, and every insight I've gained so you can take control of your health, your future, and your whole-BEING wellness.

And now, I want to leave you with this:

Keep going.

Keep growing.

Keep challenging yourself.

Because you have so much more ahead of you.

Resources

Your Quick-Start Guide to Lifelong Success

This book is designed to be a long-term resource—something you can return to repeatedly as you navigate your journey to whole-being wellness. Whether you're tapering off GLP-1s, considering starting them, or simply looking for a sustainable way to lose weight and reclaim your health, this section will help you find the strategies, insights, and tools to support you at every step.

📌 **For additional resources, visit:** *www.ironcruciblehealth.com*

Whole-BEING Empowerment Workbook

Looking for a deeper dive into your transformation? The **Whole-BEING Empowerment Workbook** expands on the chapter mini exercises in this book, providing interactive tools, guided reflections, and additional strategies to reinforce lasting change.

💡 **Get your copy here:** *www.ironcruciblehealth.com/glp1-exit-workbook*

How to Taper Off GLP-1s Without Regaining Weight

📖 **Chapter 12: Breaking Free—Strategies for Tapering Off GLP-1s**

- How to safely taper off GLP-1s with your healthcare provider
- Managing hunger and cravings naturally

- Blood sugar stabilization strategies
- Setting up systems for long-term success

💡 **Bonus Resource:** *Comprehensive Plan to Wean Off GLP-1s for Weight Loss*

🔗 www.ironcruciblehealth.com/glp1-taper-plan

How to Use Your DNA Blueprint to Personalize Your Health Journey

📖 **Chapter 6: The DNA Connection—Unlocking Your Body's Blueprint for Sustainable Weight Loss**

- How genetics impact metabolism, hunger, and weight loss
- The role of DNA in creating a personalized nutrition and fitness plan
- How to get your DNA Blueprint evaluated and use the insights effectively

📖 **Chapter 10: Fitness for Your DNA and Lifestyle**

- How your genetic profile influences your ideal exercise routine
- Strength vs. endurance training based on DNA insights

💡 **Resources:** *How to Get Your DNA Blueprint & Personalized Coaching*

🔗 www.ironcruciblehealth.com

How to Manage Hunger, Reduce Cravings, and Fuel Your Metabolism

📖 **Chapter 9: Building Your Nutrition Foundation**

- The science behind hunger and satiety

- Balancing macronutrients for sustainable energy
- Managing emotional eating and food triggers

📖 **Chapter 12: Breaking Free—Strategies for Tapering Off GLP-1s**

- Strategies to prevent hunger spikes after tapering off medication

💡 **Bonus Resource:** *GLP-1 Boosting Meal Guide & Natural Appetite Control Foods*

🔗 www.ironcruciblehealth.com/glp-boosting-guide

How to Build Habits That Last a Lifetime—Without Feeling Restricted or Overwhelmed

📖 **Chapter 5: The Seven Steps to Lasting Transformation with Neuro-Associative Conditioning (NAC)**

- Rewiring your brain for sustainable success
- How to interrupt limiting patterns and create empowering habits
- Conditioning new behaviors until they feel automatic

📖 **Chapter 13: Your Long-Term Plan—Thriving Without GLP-1s**

- Setting realistic, long-term health goals
- How to recover from a slip-up and stay on track
- Overcoming setbacks and plateaus

How to Overcome Emotional and Psychological Barriers That Hold You Back

📖 **Chapter 2: Breaking Free from the Diet Mentality**

- The psychological impact of dieting and weight loss cycles
- How to reframe your mindset for lasting success

📖 **Chapter 7: Rewiring Your Neuro-Associations for Freedom**

- Understanding the power of cravings and emotional eating
- The "Tomorrow Technique" and other urge management tools

📖 **Chapter 11: Stress, Sleep, and Whole-BEING Wellness**

- The connection between stress, inflammation, and weight gain
- Optimizing sleep for better appetite control

Where to Get Additional Support

💡 **Work with Holli & Iron Crucible Health Coaching**

🔗 *www.ironcruciblehealth.com*

- One-on-one coaching options
- Group coaching & DNA-based weight loss programs
- How to get started with a strategy session

Recommended Reading List

These books complement the insights in this book, offering deeper guidance on mindset, nutrition, movement, stress, resilience, and peak performance. Whether you're looking to strengthen your habits, optimize your metabolism, or cultivate lasting well-being, these resources will support you on your journey.

Mindset & Behavioral Change

- *When Food Is Love* – Geneen Roth
- *Women, Food, and God: An Unexpected Path to Almost Everything* – Geneen Roth
- *Atomic Habits: An Easy & Proven Way to Build Good Habits & Break Bad Ones* – James Clear
- *Awaken the Giant Within: How to Take Immediate Control of Your Mental, Emotional, Physical & Financial Destiny!* – Tony Robbins
- *The 5 AM Club: Own Your Morning. Elevate Your Life.* – Robin Sharma

Nutrition, Metabolism, & Epigenetics

- *Epigenetics and the Psychology of Weight Loss: How to Lose More Weight with Less Effort* – Francisco M. Torres MD,
- *Just One Heart: A Cardiologist's Guide to Healing, Health, and Happiness* - Jonathan Fisher M.D.

Movement, Fitness, & Whole-Body Wellness

- *Move Your DNA: Restore Your Health Through Natural Movement* – Katy Bowman
- *Built to Move: The Ten Essential Habits to Help You Move Freely and Live Fully* – Kelly Starrett & Juliet Starrett

Resilience & Peak Performance

- *1% Better: Reaching My Full Potential and How You Can Too* – Chris Nikic
- *The Mountain Is You: Transforming Self-Sabotage Into Self-Mastery* – Brianna Weist
- *Life Force: How New Breakthroughs in Precision Medicine Can Transform the Quality of Your Life & Those You Love* – Tony Robbins, Peter H. Diamandis, MD, & Robert Hariri, MD, PhD

Ways to Stay Connected

Website: *www.ironcruciblehealth.com*

Join the Email List: *www.ironcruciblehealth.com/glp-1-freedom*

Book a Consultation: *https://IronCrucibleHealthScheduling.as.me/*

Share Your Story: Tag me @tri_holli (Instagram), or @hbradishlane (Facebook) or use #IronCrucibleHealth

Leave a Review: If this book has supported your journey, I'd be honored if you'd leave a review on Amazon or wherever you purchased your copy. Your feedback helps others find this message and take their own first step.

***

With Gratitude

To God—the source of every opportunity, every lesson, and every ounce of strength I've ever had. None of this would be possible without Your divine guidance, and for that, I am eternally grateful.

To my husband—your unwavering support has carried me through every challenge, every leap of faith, and every far-out dream I've chased. I couldn't do this without you.

To my clients—you are the heart of this work. Your courage, resilience, and commitment to transformation inspire me every day. Thank you for trusting me with your journey.

To my early readers — **Dale Amory, Joyce Coffey, Alan King, Robert Lane, Janice Lowstuter, and Anne Poole** — thank you for your honesty, encouragement, and thoughtful insights. Your feedback helped shape this book into something more powerful than I could have created alone.

To my editor, Susan Haswell-Keillor — thank you for helping me clarify my voice while staying true to my message. Your guidance brought strength and refinement to every page.

To Dr. Pamela Buchanan—thank you for your generous endorsement, clinical insight, and the heart you bring to this space. Your contribution added both depth and credibility to this work, and I'm truly grateful for your support.

And to you, the reader—you've already taken a powerful step. Recognizing the need for change and seeking the tools to reclaim your health is no small feat. Empowerment begins with a decision, and you've made it. I commend you for showing up, for choosing growth, and for investing in a future where you are truly thriving.

The longer I live, the more I see that gratitude is the language of a full life. *Thank you.*

About the Author

Holli Bradish-Lane is a licensed respiratory therapist, DNA-certified health coach, certified personal trainer, and former healthcare leader with over two decades of experience in Quality, Regulatory, Patient Safety, and Performance Improvement. As a pioneer in DNA-based health coaching, she helps individuals taper off GLP-1 medications using personalized, science-backed strategies for lasting weight loss and metabolic wellness. Holli is redefining the path to sustainable health—empowering people to break free from dependency, reclaim their bodies, and thrive for life.

But beyond credentials, Holli is a storyteller, a seeker, and a soul rooted in resilience. She lives off-grid on a 40-acre ranch, where nature, stillness, and the wisdom of the earth inform her work. Her writing blends science with spirit—inviting others to live more present, more grounded, and more fully awake.

Final Thoughts, Stay Connected, and Keep Thriving

This book may be ending, but your journey with me doesn't have to. If you're ready to go deeper—whether that's refining your nutrition, optimizing your fitness, or continuing to work with DNA-based coaching—I'd love to be part of your next chapter.

You can connect with me through my website, explore my coaching programs, or join a community of like-minded individuals who are all committed to thriving for life.

To take the next step, you're invited to:

- **Join my email list** for inspiration, tools, and real-life strategies
- **Download free resources** to support your next phase of growth
- **Book a consultation** to explore personalized coaching
- **Leave a review** and help others find this message
- **Share your story** on social or with someone who needs to know they're not alone

Because this isn't just about losing weight.

It's about living fully.

And you? You're just getting started.

With gratitude and belief in you,
Holli

Scan to keep going →

Explore coaching & next steps

www.ironcruciblehealth.com/glp-1-freedom

About the Publisher

This book is published by Refiner's Forge Publishing — where transformation is forged, not forced.

www.ingramcontent.com/pod-product-compliance
Lightning Source LLC
Chambersburg PA
CBHW060452030426
42337CB00015B/1565